# PIVOTAL COUNTRIES,
# ALTERNATE FUTURES

# Pivotal Countries, Alternate Futures

*Using Scenarios to Manage American Strategy*

Michael F. Oppenheimer

# OXFORD

UNIVERSITY PRESS

Oxford University Press is a department of the University of
Oxford. It furthers the University's objective of excellence in research,
scholarship, and education by publishing worldwide.

Oxford    New York

Auckland    Cape Town    Dar es Salaam    Hong Kong    Karachi
Kuala Lumpur    Madrid    Melbourne    Mexico City    Nairobi
New Delhi    Shanghai    Taipei    Toronto

With offices in

Argentina    Austria    Brazil    Chile    Czech Republic    France    Greece
Guatemala    Hungary    Italy    Japan    Poland    Portugal    Singapore
South Korea    Switzerland    Thailand    Turkey    Ukraine    Vietnam

Oxford is a registered trademark of Oxford University Press
in the UK and certain other countries.

Published in the United States of America by
Oxford University Press
198 Madison Avenue, New York, NY 10016

Cataloging-in-Publication Data is on file at the Library of Congress.
ISBNs: 978–0–19–939710–5 (paperback); 978–0–19–939709–9 (hardcover)

3 5 7 9 8 6 4 2
Printed in the United States of America
on acid-free paper

*This book is dedicated to my wife Donna,
and to my three sons, Harry, Julius, and Peter.*

# CONTENTS

# ACKNOWLEDGMENTS

This book represents the accumulated insights of many years of thinking, advising, and teaching about the future; first at a path-breaking consulting company called the Futures Group, founded and presided over by Ted Gordon; then at my own consulting company, Global Scenarios, working principally on behalf of the National Intelligence Council and several think tanks; and currently as professor at the Center for Global Affairs (CGA) in the School of Professional Studies at New York University, where I oversee a Masters Degree concentration in International Relations Futures and have been blessed with extraordinary colleagues and students. As such, the people deserving my thanks are too numerous to list, and if I tried, I would doubtless leave several out.

So selectively, I am greatly indebted to the many graduate students at CGA who worked closely with me on the Scenario Initiative, an alternate futures study of several countries pivotal to US interests. For the Iraq post-2010 and the Iran 2015 studies, Ethan Cramer-Flood and Chris Janiec provided essential support; for China 2020, Bernadette Shaw wrote the scenario and the drivers paper reproduced in this book; for Russia 2020, Eduard Berlin wrote the scenario in this book and the notes on the Russia scenario discussion, and Jessica Bissett provided additional support; for Turkey 2020, Samuel Mueller and Melanie Vogelbach helped to organize the scenario workshop and compose the alternate scenario narratives; for Ukraine 2020, Bianca Gebelin and Gordon Little were key authors and organizers; for Pakistan 2020, Rorry Daniels, Regina Joseph, Ante Basic, and Aima Raza did much of the organizing and drafting; for Syria 2018, I was supported by Ashley Beiser, Jacob Kennedy, Michael Lumbers, and Gregory Viola.

I also owe many thanks to the program administrators who keep CGA running, and who organized, with barely a hitch, multiple two-day conferences at our headquarters in the Woolworth Building, bringing on six separate occasions twenty or so top experts and policymakers from all over the world to build alternate scenarios for pivotal countries. They include Cori Epstein, the Director of our master's program in Global Studies; Alice Eckstein, the Executive Director of the Center; and Anna Mosher, our Program Administrator. Perhaps their most heroic deed was to organize a workshop in the middle of one of New York's worst snowstorms on record, in the winter of 2009–10, appropriately on the future of Russia.

None of my recent work on alternate futures would have been possible without the generous support of the Carnegie Corporation of New York, and specifically of Program Administrator Patricia Moore Nicholas. Our first two scenario studies, on Iraq and Iran, were supported by Malcolm Weiner. All the above-referenced work, lists of scenario participants and CGA staff, and a description of the scenario process can be found on the website of the Scenario Initiative, www.cgascenarios.wordpress.com.

Finally, Divisional Dean Vera Jelinek, the force of nature who founded CGA and its master's program and presides over our growing student body and faculty, has supported my scenario work and our new academic concentration in International Relations Futures. She has my gratitude.

Portions of chapters 1 and 2 are drawn and revised from my piece in *SAIS Review*, "From Prediction to Observation," in the Winter–Spring 2012 issue; my thanks to the *Review*'s editors and publishers for their permission to use this material.

# PIVOTAL COUNTRIES, ALTERNATE FUTURES

# Introduction

As I write this in the spring of 2015, the Obama administration's grand strategy—combining great power cooperation, restraint in the use of force, and pivot toward Asia—is imperiled by Russia's "nineteenth-century behavior" and Middle East–wide sectarian strife. Iraq is fragmenting, succumbing to official sectarianism and the spillover from civil war in Syria. Out of Iraq in 2011, US forces have been reinserted, under chaotic conditions and without the legal protections the absence of which led to the original departure. In so doing, the administration finds itself acting in parallel with its principal regional adversary—Iran—to preserve Iraqi "sovereignty," even as Iran consolidates its influence inside Iraq. It has implicitly acquiesced in Assad's survival in Syria, issuing calls for a negotiated solution to the Syrian civil war without establishing leverage over the terms of that settlement. The "reset" with Russia has become a bad joke, hopes for a G-2 with China a distant memory. Recent events—of palpable surprise to decision-makers—may not produce a reversal of global strategy over the short term, but indicate diminishing returns to "restraint" and very likely a reversion to more robust use of US power, perhaps by President Obama himself, more likely by whoever—regardless of party—succeeds him.

Restraint and rebalancing may be fully warranted by America's resource limits, last decade's experience with the use of force in Iraq, and the shift in the world's center of gravity toward Asia. But the administration's management of the strategy has been sadly lacking. Any strategy, especially in a world of diffusing power, requires choices about where to allocate limited resources, and where not. Each of these choices is premised on assumptions about the multiple claimants on our capacity, and these assumptions, correct as they may be when made, in a fast-moving world are subject to rapid obsolescence. Risk is inherent in any strategy, and failure to recognize, anticipate, and manage risk guarantees unpleasant surprises and invites strategic failure. This applies as much to a policy of restraint as it does to one of primacy. Risk avoidance is self-deception, deferring hard choices as options for risk management disappear. Paradoxical perhaps, but essential to a sustainable strategy, is imagining how it might fail, then monitoring, preventing, and/or mitigating these sources of failure, even if this requires short-term actions, including in Obama's case the more robust use of force, at apparent variance to the overall approach. A combination of strategic coherence and tactical agility is essential. Making strategy work, in a world of accelerating change, requires alertness and adaptability, if the broad purposes of the strategy are to be upheld. This type of strategic management has been lacking in the administration, and this—rather than the strategy itself—may prove fatal to the Obama experiment with restraint.

Incompetent strategic management is a bipartisan phenomenon. The Bush administration came into office with a strong sense only of what it wanted to avoid (nation building), was then surprised by an event it might, in broad outlines, have anticipated, and embarked on a strategy of forceful, unilateral (if necessary) intervention to prevent "gathering" threats and extend democracy. It misapplied these ideas in Iraq, spent its remaining years in office trying to clean up after itself, then handed off the remaining mess to President Obama. By applying prevention in the wrong place, based on falsified intelligence, then underresourcing the effort, it did both us and Iraq more harm than good, and devalued an approach that might have proven useful to succeeding administrations facing diffuse and undeterrable threats. Neither administration seriously considered how to sustain its own strategy by managing downsides and applying resources to mitigate risk. Both substituted wishful thinking for close observation and testing of assumptions against emergent reality. In failing to manage risk, both administrations delegitimized their own strategy, and have left us still groping for the right balance between assertiveness and restraint. The failures of both administrations are now depressingly evident in Iraq.

The importance of anticipation, continuous reevaluation, and adjustment of strategy—of strategic management—was fully understood by those decision-makers "present at the creation" of American postwar grand strategy (a term that can be applied here without irony). Thus, Dean Acheson, explaining George Marshall's conception of the task faced by the State Department's new Office of Policy Planning: "to look ahead, not into the distant future, but beyond the vision of the operating officers caught in the smoke and crises of current battle; far enough ahead to see the emerging form of things to come and outline what should be done to meet or anticipate them. In doing this the staff should also do some-thing else—constantly reappraising what was being done. General Marshall was acutely aware that policies acquired their own momentum and went on after the reasons that inspired them had ceased."[1] And indeed, American postwar strategy reflected major adjustments to emergent reality (the extent of Europe's political and economic weaknesses, the rise of the Soviet threat, the need to harness German power) that brought the West from the Atlantic Charter in 1941 to the "dual containment" of liberal order building plus NATO, without losing the continuity of our commitment to full American engagement.

This book proposes the use of alternate future scenarios to help make better decisions in real time: in James Steinberg's words, "to bring the future into the present."[2] It considers alternate scenarios as a critical element in strategic management, part of an ongoing process of reality testing, observation and learning, policy adjust-ments, and, if necessary, fundamental rethinks. Open minds, early warning, and preparation for an altered environment are the purposes of the scenario process; not prediction, not planning in the sense of elaborate road maps into the future. I am fully aware of how hard this is, in a dangerous, rapidly changing, and uncertain environment, in which US decisions are both fateful and contingent. Protecting our many interests in a complex world of diffusing power demands big ideas and political commitment, but also close observation and agile adjustments. These attributes are hard to reconcile, and the balance between them shifts constantly. Debates about grand strategy, in our highly politicized environment, reward rhetorical skill and categorical pronouncements about the emerging world and our place in it. These attributes are essential for effective advocacy, but are the enemy of strategic management, which requires a healthy dose of self-criticism and adaptability to changing circumstances. It's hard to shift from one to the other, to find the right balance between commitment to a view of the world and of our interests, and pragmatism. The use of alternate scenarios, as a way of thinking and acting strategically, is an effort to get the ever-changing balance right.

The scenario process, long popular among corporate planners, has generated increasing international attention among policymakers and their analytical supporters, as complexity, surprise, and uncertainty have become the backdrop for strategic decisions in government and business. With so many key factors in motion, from the stability of the global economy to the ability of even powerful governments to maintain security in an age of globalization, interest in analytical approaches that incorporate uncertainty and rapid change is very high. The irony is that this complexity makes such techniques both more important and more difficult to accomplish, as decision-makers and observers, in and out of government, become easily overwhelmed by the noise, preoccupied with the immediate, and distracted from strategic questions. Generating real value and sustaining legitimacy for such approaches is an uphill climb, conceivable only as part of a process that enjoys top-level support. Yet the external conditions that demand such an approach are clear: a fast-moving world with multiple surprises, policy errors stemming from ill-conceived assumptions, unanticipated blowback from previous actions, and shrinking resources that place a premium on improved intelligence and effective policy deliberation.

Alternate scenarios are designed as plausible (not necessarily probable) narratives describing how very different futures could emerge from current circumstances, with markedly different consequences for US interests and policies. They thus serve to expose the alternate paths already discernable—but often overlooked—in the present, sharpen debate about prevailing trends, and reveal the limited shelf life of extant conventional wisdom. With the plausibility of distinctive futures established, they can serve as alternate platforms for evaluating the sustainability of current policies and for testing the effectiveness of new approaches.

Alternate scenarios can be organized around issues (terrorism, environmental degradation, global crime), trends (future of the Internet, alternate paths for globalization), or geography (countries or regions). The geographic focus taken in this book reflects the conviction that we are still living in a world of states (strong or weak), where power (size, resources, location, political resilience, economic efficiency, military strength) and interest are the principal determinants of state action. Territory still matters greatly. But state actors operate in a context of strengthening nonterritorial forces, of economic integration, technology diffusion, flows of ideas and information. These forces are critical in shaping the world and have their largest impact and clearest manifestation as they interact with states, reinforcing or undermining the capacity of state actors and rearranging the interstate balance of power. The interaction of the Westphalian system of "sovereign" states with transnational forces is the fundamental source of surprise and discontinuity in today's and tomorrow's world. The geographic focus also aligns scenario results with the organization of the policy process, which emphasizes states and regions, thus maximizing the value for policy reviews and decisions.

Much of the material for this book was generated during a multiyear project I directed at the Center for Global Affairs, in the School of Professional Studies at New York University. The Center, with help from Malcolm Weiner, funded scenario studies on Iraq and Iran; the Carnegie Corporation of New York supported our studies of China, Russia, Turkey, Ukraine, Pakistan, and Syria. The future of these countries is both important to US interests and highly variable. Some are potential or actual peer rivals. Others are regional competitors with sufficient capability to thwart or reinforce US objectives. Others are vectors of conflict. All are subject to internal political, economic, and environmental challenges, and to external shocks. How they respond to these stresses is consequential for US interests and policies. Policymaking for these countries therefore requires anticipation, but will not allow simple extrapolations from recent history. Alternate scenario approaches are designed for these circumstances.

This is a "how to" book, laying out the process of scenario construction in chapter 4. But process must be tailored to function (chapter 3 on the value proposition), and function must reflect the nature of the problem (chapters 1 and 2 on sources and consequences of uncertainty). The book then proceeds to suggest several challenges to US interests and policies that demand an alternate scenario approach, and discusses how to integrate alternate scenarios into the policy process (chapter 5).

# Chapter 1

# The Problem

## *Fateful Decisions in Uncertainty*

If the Cold War was defined by rigid bipolar alliances, national economies, and strategic continuity, the more recent past exhibits rapid change, more fluid alignments, increasing globalization, wide policy choice, and strategic surprise. Indeed, surprise defines the chronology of the post–Cold War period, beginning with the collapse of the Soviet Union and its empire, then running through the outbreak of ethnic conflict in the Balkans, the Mexican and Asian financial crises, the magnitude of genocide in Rwanda, Indian and Pakistani nuclear brinkmanship, global trade meltdowns in Seattle and Cancun that presaged the failure of the Doha Round, the Arab Spring, the global financial and economic crisis,[1] recent Russian policy choices that risk global isolation for modest territorial gain, Assad's resurgence in Syria, the emergence of ISIS (the Islamic State of Iraq and Syria), and the rapid disintegration of Iraq.

Surprises are to be expected. They reflect traditional forces in world politics—rising powers, emergent technology, sudden leadership change—as well as new and poorly understood factors such as physical changes to our planet, economic globalization, the empowerment of nonstate transnational actors, the transformation of ideas about politics and society, and the spread of democracy and pseudodemocracy. We are now observing diminished political authority and a weakening of previously robust global institutions under conditions of extended economic weakness, a potential "new normal" that will continue to supply abundant political stress and reconfigure relative power among states. The longer this period of economic insecurity, the greater the chances of further disintegration, particularly among states—both democratic and not—and institutions with limited popular legitimacy. These stresses are clearly evident in the EU, as they have been in the revolts of the Arab Spring, and in American populism and our polarized politics. All of this demands new tools in determining policy: We will have to deconstruct much of what we've learned about international politics during the last fifty years, if we are to successfully anticipate, and respond to, further unraveling.

Sustained pre-financial-crisis prosperity could return, fueled by cheap energy, continued productivity gains from a ubiquitous Internet, sustained growth and increased openness in Asia, and economic dynamism in Africa. Such developments could relieve some of the pressure on globalization-friendly states, even as they further marginalize states resisting these forces. But if we look beyond the still lingering economic weakness, there are several reasons to expect a longer-term future of uncertainty, rapid and unpredictable change, and frequent surprise.

## A Global Most Likely Scenario

For one thing, the proliferation in the sheer number of state actors with influence over global conditions affecting US interests, and the complex, dynamic nature of interactions among them, greatly increases unpredictability and risk. That we will continue to experience a global diffusion of power—to both other states and nonstates—is among the most widely shared predictions of international relations observers and practitioners, from the National Intelligence Council to international relations scholars of both realist and liberal orientation, to President Obama. One of neorealism's most powerful axioms is the inherently greater degree of uncertainty and conflict under conditions of multipolarity than under bipolar (the Cold War) or unipolar (the 1990s) conditions. Kenneth Waltz, the father of neorealism, explains that "in a multipolar world, dangers are diffused, responsibilities unclear, and definitions of vital interests easily obscured. ... When there are several possible enemies, unity of action among them is difficult to achieve."[2] In multipolar conditions, the greater number of consequential actors multiplies the events and policy shifts relevant to our interests. The balance of power can shift rapidly and dramatically, as coalitions break apart and reassemble. An awareness of the ever-present potential for adverse changes in the balance of power encourages states to adopt worst-case assumptions about both allies and adversaries, tempts states to preempt potential challengers, and limits the time available for careful deliberation and crisis management. Those responsible for intelligence and threat assessment are spread thinly among a multitude of potential threats, often from states whose internal politics, capabilities, and intentions are opaque, thus guaranteeing a constant stream of surprises.

A second factor is what Samuel Huntington called the non-westernization of world politics, a world of rising states with non-Western identities, interacting with increasing frequency and intensity across civilizational boundaries, producing heightened friction and conflict. These states, Huntington argued, reject Western values and institutions. "International relations, historically a game played out within Western Civilization, will increasingly be de-Westernized and become a game in which non-Western civilizations are actors and not simply objects; successful political, security and economic international institutions are more likely to develop within civilizations than across civilizations; conflicts between groups in different civilizations will be more frequent, more sustained and more violent ... the paramount axis of world politics will be the relations between "the West and the Rest. ... ."[3]

Multipolarity among like-minded states is manageable through institutions, ad hoc cooperation, effective communication, and shared values that rule out escalation of differences toward conflict and violence. Civilizational cleavages under hegemony produce inconsequential skirmishes on the periphery. Multipolarity plus civilizational conflict places incompatible identities at the center of world politics. They also greatly increase unpredictability, as threats bubble up from the internal politics of a multitude of actors, whose motivations and capabilities are shrouded in mystery, beyond the capacity of even well-developed intelligence communities to understand, much less anticipate.

Third, there is the reality of globalization, by which I mean the huge increase in flows of goods, services, investment, information, technology, ideas, and people across national borders. The widely accepted imperative of economic growth, the number of public and private players committed to globalization, and the rapid innovation and diffusion of technology guarantee continued expansion of global flows. Globalization has enabled the rise of new peer competitors to American hegemony, as flows of investment and technology have stimulated growth and modernization in low-cost emergent countries. It has facilitated the proliferation of new transnational threats, as actors from states to groups to individuals have been empowered to seek their objectives—benign or dangerous—through porous state borders. It has brought cultures of widely divergent values and identities into close proximity, generating friction and increasing mutual vulnerability. The combination of globalization and recent, extended economic underperformance has stress-tested governments and intergovernmental institutions premised on prosperity and American liberal hegemony, and some have proven too fragile to survive. Governments that lack the legitimacy to ride out global volatility have collapsed, and great powers in multipolar competition have been uninterested or unable to act collectively to prevent or contain the unraveling. Indeed, in the world as envisioned here, these bottom-up conflicts of state implosion will aggravate great power rivalry. Globalization will interact with multipolarity and civilizational conflict to produce the chaotic, unpredictable, dangerous world that we will inhabit long into the future.

Although globalization is a reasonable assumption to make about the future, the liberal version at the center of American postwar grand strategy is not. We are now observing globalization without American primacy, and it is very much a mixed blessing, delivering growth and prosperity to many, but without the strong rule-based institutions that anchored the efficient, inclusive, and reciprocal system and delivered both openness and international accountability to liberal norms. Globalization in a system of multiple, competitive actors and conflicting identities will subordinate the global economy to global politics, reducing efficiency and increasing volatility and uncertainty. Illiberal globalization will exhibit a widening range of homegrown growth models and of trade and financial practice, a proliferation of regional preferential agreements motivated more by strategic competition than by economic efficiency, and an extension of "geonomic" competition into trade practice (for imports and exports), investment regimes, currency pricing, resource access, global information flows, and technology acquisition.

This system of fractious, multipolar globalization is not premised on American decline. Assuming that we are prepared to acknowledge new realties, we have a great deal going for us in such a world, and many reasons for optimism about the future of our economy, our military strength, the soft power appeal of our system of government, our resilience in the face of shocks. Demand for American power is still high, especially from states in the shadow of rising powers. But rivals to American hegemony there will certainly be, and their power and ambition will continuously chip away at our regional primacy in Asia, Europe, and the Middle East, as they challenge the liberal rules of the game in trade, technology, and access to the commons. The future will reflect our preferences to a much lesser extent than we are accustomed, as a multitude of disparate and conflicting actors vie for influence within their regions. Local balances of power will shift rapidly; threats will appear seemingly out of nowhere. And while our ability to "create our own reality" will be diminished, our interest in distant events will remain. If, as I expect, the next president tacks back toward primacy as a "grand strategy," we will find the world less receptive than previously, more motivated and able to actively resist, in the case of rivals, or opt out, in the case of allies, and the result will be more state-to-state competition and less cooperation.

In our scenarios work at CGA/NYU, we applied these propositions at the country level in our studies of Russia (February 2010) and China (October 2009).[4] Could we construct plausible scenarios that delivered economic growth sufficient to sustain their challenge to US hegemony, without the liberal economic and political reforms that Americans view as essential to modernization, and which would mute their challenge to our interests? Reproduced below are one of three scenarios constructed for each country. They do not constitute forecasts. Their purpose is to raise important questions about the presumed constraining effects of globalization on our rivals, and the future ability of the United States to leverage globalization through economic sanctions, in order to moderate Russian or Chinese behavior. With respect to the ongoing conflict in Ukraine, the Russian "Working Authoritarian" scenario asks whether a revisionist Russia is limited to China in its search for willing external economic and financial partners, or—in a multipolar world—has European and other Asian options as well.

This particular scenario for Russia cannot be described as "accurate," at least through 2015. Indeed, the Russian "Degeneration" scenario, done in the same exercise, parallels recent developments in Russian politics and economics more closely. It does suggest the possibility, however, that if Putin curbs his appetite for further land grabs in his near abroad, the memory of Crimea and eastern Ukraine will yield to European weakness and economic self-interest, and that the erosion of sanctions, global economic recovery, and rebounding energy prices will restore the viability of "working authoritarianism" by 2020. If not, the scenario may enable us to conclude that China possesses the political and economic strengths to reconcile authoritarianism with economic modernization, but that Russia does not.

## RUSSIA: *WORKING AUTHORITARIANISM*

Coming out of the economic downturn, Putin is under public pressure to improve economic conditions and the quality of life in Russia. It becomes clear that Russia's current economic model will not provide the basis for future growth or the ability to meet the public's expectations. Efforts to modernize Russia's industrial base face overwhelming odds of failure. Even Russia's energy sector, the traditional engine of growth, is mired in inefficiency and corruption. Russia's subordinate role in the global economy has led to feelings of anger and resentment.

In this scenario, conflict within the upper echelons of the Moscow elite becomes the political catalyst behind Putin's decision to recalibrate domestic and foreign policies. He launches a radical program for achieving global competitiveness. The combination of restrictive trade barriers, deepening economic engagement with Germany, China, and South Korea, large-scale foreign investment, and the restructuring of key industries collectively results in the resurgence of Russia's economy. Russia "opens" its energy sector and other key markets to foreign participation, but on a highly selective basis. China, Germany, and South Korea become strategic stakeholders in Russia's economic rebound. All three possess capital and technology and do not require political reform as a condition for their participation in these ventures.

The key to Putin's success is his ability to leverage strategic relationships and to modernize Russia's industrial base without relinquishing his grip on political power inside Russia.

## Drivers of This Scenario

The working authoritarianism scenario emerges due to the cumulative effect of the following drivers:

### Domestic Politics

No serious opposition party emerged to contest United Russia's control of electoral politics in Russia. Instead, competition emerged from within the ruling elite and forced Putin to embark upon a radical economic recovery plan. Putin's popularity within Russia declined after the 2012 election, but public dissent is repressed. Putin and United Russia's political fortunes rebounded with the improving economy toward the end of the decade.

### Global Economic Trends

The global economy succumbed to a double-dip recession early in the decade. Lingering solvency issues in the European Union reduced Russia's ability to tap global capital markets and further weakened the ruble. After a partial economic rebound in 2010, Russian annual GDP growth flattened to 0 percent by 2012. Highly restrictive trade policies were implemented to shield Russia from foreign competition. Income from new import taxes enabled the government to fund a portion of its domestic programs and balance its budget.

### Diversification and Economic Reform

Weakening global demand for oil and gas masked the deep structural problems in Russia's energy sector. However, addressing these structural problems and increasing competitiveness would require substantial capital investment. Rather than reform the domestic economy on the "Western model," the Russian government elected to take a more mercantilistic approach, erecting nontariff trade barriers and negotiating special commercial alliances with foreign governments. Consequently, Russia gained access to capital and technology without liberalizing, suggesting that it could decouple its economic growth and political reform, at least temporarily.

*Energy Markets and Resource Management*

Russia's unproductive and inefficient energy sector faced serious structural problems that had consistently been masked by high global demand. The introduction of competitive technologies and new energy extraction processes threatened the oligopolistic positions of energy-exporting countries and their profitability. By the end of the decade, Russia streamlined its energy sector and gained access to the investment capital necessary to improve its competitive position globally.

*Foreign Policy*

With its economy in crisis, Russia faced the difficult choice of either becoming a junior partner to China or accepting second-class status as part of an enlarged Europe. By reaching out to both, Russia was able to avoid dependence on either, while continuing to reject Western liberal institutions. As its strategic partnerships began to generate investment and growth, Russia began to dominate the Commonwealth of Independent States (CIS) commercial and economic sphere.

## The Path to 2020

### 2010–2011: Continuing Economic Stagnation

Just as Russia was beginning to rebound from the global economic crisis, it found itself confronting a series of new economic challenges. Oil prices reached US$90 per barrel in the spring of 2010, but by early 2011, they dropped back to the US$60–US$70 range. By the middle of the year, the Greek sovereign debt crisis had turned into a full-blown eurozone crisis,[5] which, coupled with China's policy of economic "cooling off," triggered a second global market collapse and a predictable reduction in the demand for energy resources. Russian exports declined precipitously, affecting the entire Russian economy. GDP growth projections for 2011 were cut back from 4.5 percent to under 1 percent, well below previous forecasts. While official unemployment numbers dipped below 10 percent, the actual jobless rate was estimated to exceed 16 percent. Russia's balance-of-payments surplus was halved, driving down the ruble and putting upward pressure on inflation. Lower-than-forecast government revenues forced Moscow to reduce vital social programs and delay highly visible regional infrastructure projects.

Concerns regarding European bank solvency persisted well into 2011, just as Russia was reentering the global capital market. The widening yield curve drove up public- and private-sector borrowing costs, which only compounded Russia's problems. Contrary to President Medvedev's soothing pronouncements that Russia had weathered the worst of the recession, all was not well on the economic front.

The crisis in Russia's energy and extraction sector was particularly troubling. Forecasters attributed the sharp reduction in domestic energy production to multiple factors: lower yields from maturing oil fields, poor pipeline maintenance, project delays, and lack of refinement capacity. No new fields had been put into production for several decades. The combination of production inefficiencies, choking bureaucracy, endemic corruption, and a history of indirect appropriation of foreign assets (such as in the TNK-BP dispute) constrained Russia's ability to attract the new capital necessary to address problems in its energy sector. In addition, Russia, as with all energy-exporting countries, faced a downward market trajectory due to the introduction of liquid natural gas and shale gas and the long-awaited flow of Iraqi oil onto global markets.[6] Russian policymakers had no immediate cure for the chronic inefficiency and lack of managerial competence found in the majority of Russia's state-controlled enterprises. Gazprom was emblematic of the overall problem. The majority of Gazprom's capital expansion budget was tied up in new pipeline distribution projects in Europe, rather than in securing long-term gas supplies.[7] These projects were driven by political objectives, and many were not financially viable. One prime example was the South Stream pipeline, launched primarily to deter European countries from pursuing alternative sources of natural gas (especially Nabucco), to which Gazprom committed in excess of US$20 billion, weakening its balance sheet and raising concerns about its liquidity.

With its domestic sources of oil and natural gas beginning to dwindle, Russia was forced to rely increasingly on Kazakhstan and other Central Asian countries to meet Europe's demand for oil and natural gas. The need to backfill supply from foreign sources reduced the profitability of Gazprom and of Rosneft, Russia's state-controlled petroleum enterprises. It was becoming clear to Russian leaders that it would take more than a rise in global energy demand to address the problems facing Russia's carbon-based economy.

The Medvedev-Putin administration faced a number of other domestic problems. Corruption was spreading. New accounts of graft and extortion in high places were being reported on a weekly basis. The public expressed outrage over the high cost of imported food, household electronics, and automobiles. The "second dip" in the economy had triggered a new wave of capital flight, such that the financial system was once again on the brink of failure and required a massive infusion of public funding. Higher prices, bank failures, and official corruption bred deep resentment and spurred larger and more frequent public demonstrations, even in Moscow and St. Petersburg. Nearly all nonenergy sectors of the economy were suffering setbacks. General Motors' decision to retain its controlling stake in the Opel automotive operations in Germany, rather than selling to a Canadian-Russian investment consortium, was a major setback in Putin's efforts to resuscitate Russia's automotive industry.[8] Chinese munitions manufacturers had made headway into Russia's lucrative arms export market. Turkish, German, and Chinese engineering firms had been the main beneficiaries of the decade-long construction boom in Russia, bypassing domestic suppliers.[9] Well-reported plans to establish a domestic pharmaceutical industry remained on hold due to the lack of capital, and foreign pharmaceutical companies remained skittish about doing business in Russia in the absence of adequate patent and copyright protection. Even the much-touted nanotechnology initiative was at best a long-term strategy.[10]

The Putin-Medvedev "tandemocracy" deflected blame for the stream of bad news and public disenchantment because, quite simply, there were no political alternatives on the horizon. The Kremlin continued to undermine opposition parties, and, with the help of an army of marketing and public relations specialists, distracted a weary public by flooding the Russian media with items about the upcoming Olympics, improving relations with Ukraine, and the soon-to-be realized oil riches from the Arctic. Even skeptical members of the press found themselves spending more time focusing on the possible political showdown between Medvedev and Putin in the upcoming elections than on the financial plight of Russian households.

## 2011–2012: A New Paradigm

Medvedev, Putin, and their advisers recognized that there were no easy solutions to Russia's economic malaise. Although the Kremlin understood the criticality of addressing structural problems in the economy, its primary objective was to retain its political authority. Putin and his *siloviki* supporters summarily rejected the liberal-oriented economic solutions proposed by Igor Yurgents and his reform-minded advisers in late 2009, which stressed the importance of an independent judiciary and enhanced property and contract guarantees, steps that could lead to calls for more political openness.[11]

Putin knew that Russia was facing a long-term downward economic trend and that it would likely need financial assistance from the West, but he was also aware that global investors and multilateral institutions would insist on painful and politically destabilizing reforms as prerequisites to the provision of capital. Russian leaders continued to object to efforts by the EU and the United States (via the International Monetary Fund) to make future economic dealings with Russia contingent on its willingness to commit to political liberalization.[12] They also feared becoming economically dependent on China.[13] Putin recognized the importance of balancing China's growing economic dominance in Northeast Asia.

While Medvedev and Putin publicly advertised Russia's economic potential, a debate surrounding the future direction of the country raged in the Kremlin. Younger members of the Russian elite, who had missed the chance to participate in the privatization schemes of the 1990s, demanded the Kremlin shift direction, but were unable to propose a workable alternative. Older statesmen, many of whom have been the major beneficiaries of the graft-ridden system, insisted that state control of the economy was critical to Russia's long-term global competitiveness. Putin found himself caught in the middle of an internecine struggle between the oligarchic crony capitalists and a coalition comprised of key members of the *siloviki*, Western-trained technocrats, entrepreneurs, and nationalist elements within United Russia, all of whom felt humiliated by Russia's economic circumstances and resentful of Russian billionaires, who lived and played abroad while Russia faced such problems.

Gazprom and its leaders became the major target of criticism. To its detractors, Gazprom represented everything wrong with Russian industry: it was bloated, inefficient, and poorly managed. Dmitri Medvedev, a former darling in reform-minded circles, became the scapegoat responsible for decimating Gazprom's balance sheet and whitewashing its dismal project track record during his tenure as Gazprom's chairman.[14] Although Putin could just as easily have been blamed, to no one's surprise, Medvedev wound up taking the bulk of the criticism. As his political stock fell in the run-up to the 2012 presidential election, Medvedev quickly became a lame-duck president. As a result, United Russia leadership, thinking only in terms of survival, looked to Putin to rally the country and secure the Kremlin for the next six years. They beseeched Putin to run in 2012, and he relented. Medvedev offered his full support. Although United Russia had encountered problems in the most recent regional elections, Russia's opposition parties remained fractured and politically inept.[15] Thus, Putin was able to glide through the 2012 election campaign without serious opposition, and he was, once again, elected president.

It became clear from Putin's cryptic campaign speeches that there would be a massive upheaval in the near future. Within days of the election, Putin requested and received legislative approval to return executive authority to the president, thus reducing the role of the prime minister. Putin then announced the formation of his new government. He nominated Sergei Lavrov, the current Russian foreign minister, as prime minister, emphasizing the importance of foreign affairs in Russia's economic recovery program. Vladislav Surkov, Putin's longtime adviser and proponent of the centralization of government authority (i.e., the "power vertical") was responsible for designing Russia's new economic and industrial strategy. Aleksei Kudrin retained his Central Bank portfolio. Igor Sechin was given responsibility for trade relations, and Dmitry Rogozin was named ambassador to Germany.

Putin's new program included five main components and would prove a radical departure from the previous decade. First, it called for a massive restructuring of Russia's energy sector in an effort to spur innovation and professional management practices and instill a sense of urgency among Russia's energy titans. Second, broad trade restrictions were to be erected, encompassing all essential industrial sectors, such as construction materials, automotive parts, industrial chemicals, cement and asphalt, and memory chips, with the objective of reducing dependency on foreign products and improving the competitiveness of Russian firms in these sectors. Import duties and "domestic content" requirements were established, despite their prohibition under World Trade Organization guidelines. Putin, thus, categorically rejected the advice of Moscow's liberal, reform-minded think tanks. Third, the plan called for targeted demonopolization in peripheral industrial sectors to encourage both innovation and direct competition. Putin needed Russia's entrepreneurs to play a role in a technological reawakening across Russia's domestic economy and created an Industrial Investment Fund (IIF) to capitalize new business ventures. Fourth, the Kremlin placed restrictions on wages across all state-controlled enterprises and all domestic companies that did business with the Russian government or received government funding. This included practically every domestic firm with more than fifty employees. The Kremlin also accelerated the phasing out of the electricity and petrol subsidies in an effort to increase the working capital of utility companies. Lastly and most dramatically, Putin's plan called for the creation of deep commercial alliances with a number of strategically important countries, most notably China, Germany, and South Korea. These partnerships were designed principally to provide Russia with access to capital, technology, and managerial expertise. In return, China, Germany, and South Korea would receive preferential treatment from the government and would have a distinct competitive advantage in tapping into Russia's consumer markets and vast energy and mineral resources.

The Kremlin needed to prepare the country for this major policy shift. Russians had always looked to their government to manage the economy, while providing little input. Public opinion, however, had become quite important to the Kremlin. Moscow launched a number of new Internet sites extolling the virtues of the new economic strategy and promising that wage ceilings were temporary. The message to the public was that Russia had entered a new phase in its quest to become a global power. Its partnerships with key industrial nations signaled Russia's global importance and status as a key emerging market.

Building strategic alliances with China and Germany was the centerpiece of the new economic plan and was conceived as a way to reinforce Russia's status as a key emerging market. Moscow first looked to China. Russia's commercial relationship with China had deepened considerably since the Russia-China summit in 2009, during which Putin and Chinese premier Wen Jiabao reached agreement on a number of key projects, including the development of high-speed rail service between the two countries to facilitate commodity transport and the creation of a new pipeline to transport Siberian natural gas to China.[16]

Putin's one-on-one session with Chinese president Xi Jinping at the Shanghai Cooperation Organization meeting in 2012 focused on how to further expand the partnership. For example, recognizing China's need for timber due to rapid deforestation of its northern regions and Russia's need to revitalize its inefficient forestry industry,[17] Xi and Putin concluded a ten-year timber agreement, under which Russia would commit 75 percent of its timber exports to China and China would finance the modernization of dozens of mills across Siberia and the Russian Far East. Putin also agreed to provide China with up to 75,000 nonresident work permits to accommodate the need for Chinese workers to handle the large number of projects. In addition, during the meetings Xi voiced China's concerns about growing Russian protectionism, and the two countries agreed to continue to lower barriers to cross-border trade.[18]

Later that year it was announced that Russia and China had formed a new Far East Commodities Development venture partnership. This new venture, based in Vladivostok, was granted a fifty-year exclusive exploration charter to establish iron ore mining and energy distribution facilities across the Russian Far East. Russia retained a 51 percent interest, and, in return, it was guaranteed US$3 billion in annual income from the venture, plus 25 percent of all revenues derived from international sales.

The relationship between Russia and Germany was also about to expand significantly. The two countries had successfully collaborated on the development of the Nordstream gas consortium, which was expected to become the primary source of natural gas delivery to northern Europe. German investors and manufacturing firms coveted Russia's untapped mineral and energy endowments. In addition, Germany's electronics and automotive sectors sought to penetrate Russia's expanding consumer markets. Thus, Germany pursued deeper commercial links with Russia, despite criticism from fellow EU member-states.[19]

Under the terms of the new alliance, Ruhrgas, the German energy conglomerate, became a crucial stakeholder in the newly formed transnational natural gas consortium to be built upon the foundation of the proposed merger of Gazprom and its Ukrainian counterpart, Naftogaz.[20] This German-Russian-Ukrainian consortium would own 80 percent of the natural gas pipelines across the Eurasian landmass. Berlin, Moscow, and Kyiv approved the venture in late 2012, and installed former Russian president Dmitri Medvedev as CEO. The venture became the largest natural gas company in Europe and, under its new management and streamlined operational structure, immediately dominated gas distribution in Europe and the CIS. Over objections from Poland and the UK, the venture was approved by the EU's Competition Commission, despite its obvious anticompetitive aspects.

Russia's partnership with South Korea began with a series of strategic mining and semiconductor alliances. South Korea's Samsung and its Russian counterpart, Elbrus International, jointly established large-scale chip design and manufacturing facilities in Russia, which were exempt from Russian import and domestic content restrictions. The Russian–South Korean venture agreed to grant to Russian firms long-term licenses to advanced technology for use in the Russian military technology sector. In return, Samsung gained a major foothold in the Russian computer and mobile phone markets. In early 2013, Russia and South Korea created a global nuclear energy venture. Korean Power Engineering Company (a subsidiary of Korea Electrical Power Corp or KEPCO) joined with Russia's Atomenergoprom, the leading Russian supplier of nuclear power facilities, combining technologies and market presence to become one of the leading suppliers and operators of nuclear power facilities in the world. Independently, the two companies had been awarded contracts in the United Arab Emirates, Turkey, and Saudi Arabia.[21] Collectively, the new venture was expected to win 40 percent of the international projects up for competitive bid over the next decade.

Prime Minister Lavrov's primary role in the new administration was to ensure Duma approval of new legislation that elevated these new venture agreements to the status of treaties under Russian law, thus providing German and Korean companies with enhanced contractual protection and property rights. These agreements also provided these companies with competitive advantages over other foreign investors. Unlike the United States and the EU, the German and South Korean governments decided to decouple commercial relations with Russia from their human rights and democratization preferences. All references to EU normative standards were excluded from the language of the venture agreements and subsequent domestic legislation enacted in Germany.

These alliances provided the Russian government with US$20 billion in new capital over a ten-year period. A significant portion of these funds was earmarked for oil and gas exploration, refining, and distribution. The remaining capital was allocated to a new industrial investment fund to provide capital to new businesses, primarily east of the Urals and in the Russian Far East.

Once these alliances were in place, Putin moved aggressively against the Russian oligarchy. The oligarchs had much to lose and violently opposed Putin's strategy. They accused Putin of placing his trust in foreigners rather than his own countrymen, which was not far from the truth. To promote professionalism and displace entrenched oligarchs, Reutersberg and other newly appointed members of the leadership team were recruiting professional managers from leading German, Korean, and

American companies. To reinforce this new commitment, the Kremlin promulgated a set of corporate governance guidelines, including draconian penalties for mismanagement, stiff anticorruption provisions, and performance incentive bonuses.

## 2013–2014: Foreign Diversions to Deflect Domestic Woes

The dramatic shift in Russia's economic strategy stunned its population and enraged Putin's political rivals. The Brown-Red parties accused Putin of selling out Russia's national interests to foreign banks and corporations. Foreign ownership stakes in Gazprom and Russia's other energy giants led to sharp criticism of Surkov and Putin. Midlevel workers bemoaned caps on wage increases; absenteeism increased at large SOEs. Rising electricity costs hit households hard and stoked demonstrations across the country. Opposition parties complained that none of these issues had been raised during the presidential election campaign.

The government harshly repressed street demonstrations, breaking up public meetings with a brutality rarely seen in Russia since Soviet times. It closed down a number of independent newspapers and blocked access to popular websites that had allegedly published "inflammatory and libelous" accounts of the government's response to the civil disobedience. Rosneft continually laid off masses of workers for participating in "sick-outs" and distributing anti-Putin literature.

The level of repression was widely reported in Europe and the United States. Members of the European Parliament criticized Russia for its new mercantilist policies and for intolerance of public dissent and "Stalinist tactics" for managing it. The WTO formally rejected Russia's pending request for membership, citing Russia's imposition of new trade barriers.

Tension also increased along Russia's borders. Russian intelligence had advised Putin that Georgia was harboring terrorists who were planning attacks during the Sochi Olympics in the following year. In response, Russian security forces conducted raids inside Georgian territory, and Putin made dark threats of coming to Tbilisi to "clean house once and for all," which was widely interpreted as a threat to overthrow the Saakashvili government. These actions triggered emergency NATO meetings. Russian security forces were also active in Central Asia. Soon after the death of longtime Uzbek dictator Islam Karimov in November 2013, Russian forces (reportedly at the "request" of the new government in Tashkent) swept into the Fergana Valley, ostensibly to root out a group of Islamic fighters with ties to Dagestani Jihadists.[22] A battalion of Russian peacekeepers remained in Uzbekistan to deal with the "ongoing threat" to the stability of Uzbekistan and the region. Russia's efforts to consolidate its positions in the Caucasus and Central Asia alarmed the United States and the UK and drew a strong rebuke from Beijing, which feared for its own commercial interests in Central Asia. Many Western analysts believed that the Kremlin had created these provocations in an effort to divert attention away from its failing economy and the harshness of its new economic plan.

Putin's most provocative foreign initiative involved political reunification with Ukraine. Ever since the 2010 Ukrainian presidential elections, relations between the two countries had improved.[23] Russia and Ukraine continued to seek new forms of economic and political collaboration. The Ukrainian Regions Party launched a campaign in support of a formal union with Russia, and eventually legislation was introduced in the Ukrainian Lada to call for negotiations to establish a federation with Russia. Opposition parties in Ukraine, such as the All-Ukrainian Union "Fatherland," attacked these efforts as unconstitutional and warned that Russia was seeking a stranglehold over the country.

Poland and Lithuania, fearing the loss of Ukrainian sovereignty, threatened to seek economic sanctions against Russia at the EU, but these calls were deflected by Russia's major energy partner, Germany. The EU found it increasingly difficult to build consensus around a response to the possible Russia-Ukraine reunification, given the still-frayed relations that resulted from the bailouts of Greece in 2010 and Ireland in 2012 and Germany's tilt toward Russia. Many EU members criticized Germany's relationship with Russia. German leaders maintained that their approach was consistent with the EU's policy of gradually integrating Russia into the European community.[24] In Russia, these international debates boosted support for the Russian government and distracted the public from the more adverse effects of the economic plan.

By mid-2014, after nearly three years of stagnation, the economic plan began to show results. New oil-drilling projects had finally begun. Plans to modernize the Ukrainian gas pipeline network were underway. Hundreds of new commercial ventures had received funding from the Industrial Investment Fund (IIF). Thus, while the situation was fragile, Putin and his team were still in charge. United Russia retained its commanding majority in the Duma after regional elections in 2014 and held on to key posts in Russia's regions.

## 2015–2016: Signs of Emerging Economic Growth

The economy finally stabilized in mid-2015, aided by an improving global economy, higher industrial productivity, and the increasing competitiveness of Russia's non-energy-related exports. GDP growth increased to 4.5 percent in 2016. Energy economists forecast the first net increase in Russian oil and natural gas exports since 2006. Gazprom announced the cessation of the South Stream project, and redirected its focus to modernizing Ukraine's gas pipeline facilities. A spike in employment boosted the Putin administration's standing with the public, and demonstrations became smaller and less frequent. Three years of wage controls and the elimination of electricity and petrol subsidies had taken their toll on workers and households, but social programs were fully funded in the 2016 budget. In the context of improving domestic affairs, the Kremlin began to repair relations with its neighbors. Increasing demand for Russian oil and gas in 2015 enabled Russia to leverage its energy dominance in Europe once again, and it offered price concessions to a number of influential European countries. The government's announcement that it would phase out import restrictions improved relations with China. Russia's external relations appeared to be returning to normal. Moreover, foreign investment flows were increasing as European and Chinese companies entered the seemingly untapped Russian market.

For the first time in nearly a decade, there was a sense of optimism and political stability in Russia.

### 2017–2018: Momentum Leading Up to the Presidential Election

The general opinion in Russia was that Surkov's version of "shock therapy" had put Russia on the right track. Both public and private debt had been reduced. Productivity had increased due to new management and corporate governance guidelines. The most impressive sign of economic progress was the improved performance of Russia's restructured energy sector: new projects were being completed on time, incidence of industrial accidents had declined, and the new Gazprom was finally being run like a for-profit enterprise rather than a political arm of the Kremlin.

Putin's import-substitution strategy had allowed Russian companies to dominate domestic auto parts, light machinery, steel, and lumber markets. In addition, improved technological and management practices had enabled Russian companies to expand across the CIS and into select European markets. In an attempt to replicate China's successful currency management practices, the Russian Central Bank pegged the ruble to the euro, thereby enhancing Russia's regional price competitiveness. By January 2018, Russia reported the highest increase in non-energy-related exports in two decades. Hundreds of new firms that had received IIF funding were thriving and distributing their products in foreign markets. Lower-cost Russian steel and light machinery were selling well in Europe. The merger of Russian and Ukrainian steel in 2017 had created a dominant global competitor, taking market share away from China and Japan. [25]

The economic expansion had its unintended consequences. Russian subsidiaries of German and Korean enterprises, which were exempt from import tariffs and domestic content restrictions, began to dominate the domestic automotive and computer markets, thwarting Putin's objective of creating Russian champions in each of these important sectors. In addition, corruption in both government and industry was once again on the rise, threatening to undermine productivity gains. Moscow countered by requiring the frequent rotation of managers in state-owned enterprises and by loosening restrictions on non-resident work permits that would enable the hiring of managers from Europe and the United States.

Attention turned to the upcoming 2018 presidential election. Putin announced his intention to retire from public office after completion of the current term, but United Russia officials launched a nationwide "Stay Vladimir" campaign and he agreed to run for reelection in 2018. He based his campaign on a commitment to withdraw wage caps and to offer electricity rebates to households on a lottery basis. To no one's surprise, Putin received 78 percent of the votes cast.

## 2019–2020: Tension among Partners

The election results and an improving economy emboldened the Putin administration, which began to seek new ways to leverage Russia's newfound economic and political strength. Moscow initiated negotiations with Japan over Arctic transport links, including expanding the existing Sakhalin Project to increase natural gas pipeline capacity. Russia and Japan entered into an agreement to build a cross-border high-speed rail.[26] Although China had expressed its displeasure with the prospects of Japan gaining a foothold in Russian Siberia, Russia moved forward and concluded additional agreements with Japan.

The Russia-China relationship ran into further turbulence several months later. Russian Interior Ministry officials based in Khabarovsk reported that there were in excess of 250,000 Chinese laborers working and living in Russian territory across the Amur River, far exceeding limits imposed in 2012. The Russian government demanded an explanation from Beijing. Rather than responding to this accusation, Chinese officials renewed criticism of Russian restrictions on Chinese imports, arguing that Russian trade barriers were specifically targeting Chinese products. China accused Russia of deliberately delaying its entry into the World Trade Organization in order to perpetuate its protectionist policies. This dispute developed into a full-blown diplomatic crisis.

Putin and Xi resolved the standoff. Russia agreed to expand the number of Chinese guest workers allowed in the Russian northeast to 150,000, and China committed to implementing stricter border controls. Russia also lifted import duties on several dozen Chinese products. Parenthetically, Russia's WTO membership application—strongly supported in Europe and the United States—was approved in early 2020 on the basis of

Russia's commitment to phase out all import restrictions and export subsidies by 2024.

Russia's relationship with Germany was also in flux. In 2017, the Social Democrats and their Green Party partners had gained control of the upper house in the German parliament and were able to select the next German chancellor. Formerly, while in the minority, the Social Democrats had been highly critical of German policy toward Russia, citing the apparent willingness to overlook Russian human rights violations and lack of progress on electoral reform.[27] Consequently, the change in leadership in Berlin led to a chill in relations with the Kremlin and strained the German-Russian commercial alliance.

Although by 2020 the global economy had regained momentum, global energy prices had leveled off. Iran had reentered the global oil and natural gas markets after becoming a signatory of the Global Non-Proliferation Treaty and pledging to destroy its modest nuclear arsenal. This led to a significant increase in the supply of oil on the world's markets. As a result of flagging energy prices, Russia, once again, faced an economic downturn. However, because it had become far more diversified, Russia found itself less susceptible to fluctuations in the global energy markets than earlier in the decade.

## Implications for US Policy

This scenario presents a Russia that succeeds in upgrading its economy without instituting the liberal reforms Americans associate with modernization and growth. As such, it presents an inherent challenge to our mindsets and expectations: a statist approach that diversifies and expands the Russian economy, enhances Russian power and influence, and fuels a more assertive foreign policy, without succumbing to political reform or sacrificing its autonomy to a web of trade and financial interdependencies and liberal global institutions.

This direction in Russian policy is in keeping with recent trends in the way emerging countries engage the global system. Liberal global institutions have atrophied as regional and bilateral arrangements have proliferated. States have sought strategic alliances with like-minded partners having compatible economic interests. National political economies have taken on a homegrown quality, as the returns to neoliberal formulas have diminished. The continued imperative of engaging the global economy now generates a multitude of national approaches. State direction of investment and outright ownership of strategic industries, highly selective trade and financial liberalization, local content requirements, weak intellectual property protection, mercantilist pricing of currencies, are all in the policy mix. A highly volatile global economy further reinforces the movement away from liberalism.

Russia, in this scenario, is both a reflection of these trends and a contributor to their acceleration. It finds significant trade, investment, and technology opportunities with select partners who are prepared for their own reasons to operate outside the multilateral system, and to provide capital, markets, and technologies essential to modernizing and diversifying the Russian economy. These relationships—principally with China, Germany, and South Korea—are essential to Russia's success in this scenario, and allow us to think systematically about what this future might mean for US interests.

The gains from these relationships will grow Russian power and enable a more robust foreign policy, improve ties with American allies (Germany, South Korea) and potential adversaries (China), and further contribute to a diffusion of global power and a weakening of American-brokered global institutions. Given this outcome, the scenario would be viewed with some alarm in Washington. A Russia that has never abandoned its strategic preoccupation with America as principal adversary, now operating with enhanced capacity in a multipolar system with powerful partners, would indeed present a formidable challenge. The prospect of a Russia-China collaboration on a range of issues would be troubling (though Russia's threat perception could shift as Chinese power continued to grow); Germany's commitment to Europe, and possibly to NATO, would be questioned (reinforced by a visibly growing skepticism within Germany about the gains from its EU leadership); Russian assertiveness in its near-abroad would present challenges to both American and European economic and geopolitical interests, and precipitate conflicting responses from within Europe, and across the Atlantic; Russia would be better positioned to complicate the American policy agenda of global governance and economic liberalization, democratization, nuclear nonproliferation, stabilization of failing states, counterterrorism, and Middle East peace (and also to win benefits in exchange for its collaboration).

Yet the fact remains that US-Russian interests are coincident in many respects, and the challenges presented in this scenario are balanced by the potential benefits of a "successful" Russia: one that is relatively secure within its own territory, has leverage over other states and the capacity to act effectively—as adversary but also as partner—and the self-confidence to finally move beyond its sense of grievance at the collapse of the Soviet Union. The interests we share are compelling: slowing the spread of weapons of mass destruction and nuclear materials; addressing the threat of terrorism; reducing or containing the chaos of failing states; managing the rise of China; enhancing and stabilizing global growth; and confronting the (hopefully) shorter-term challenges of containing or deterring Iran and stabilizing Iraq and Afghanistan. All these issues could easily become the substance of future US-Russian conflict, and in even the best case our natural rivalry will prevent close, permanent collaboration. But joint management of some of them, and genuine collaboration on a few (as is the case with arms reductions and control of nuclear materials) should not be beyond our capacity.

In this scenario, a US-Russian "limited partnership" would openly acknowledge differences, accept sovereignty and noninterference as the basis of the relationship, operate largely through regular bilateral channels, and embrace a specific, limited agenda for cooperation: nonproliferation of WMD weapons and materials; counterterrorism; arms reductions (to enhance mutual security and lend credibility to the nonproliferation regime); bilateral trade and investment promotion (the United States can play this game as well as Germany and China); and managing China's rise. Off the table are human rights, WTO accession, and NATO membership (probably no harm in offering closer association, but less reason here for Russian interest). Stabilizing and growing the global economy might be included, centering on institutionalizing the G-20 and reforming the IMF to enlarge emerging country representation. Middle East peace is worth trying, given the benefits for containing terrorism and the spread of nuclear weapons, but here the temptations for each country to seek special advantages require limited expectations.

Realizing the upside of this scenario requires that both sides see and act on these possibilities. For a successful Russia, the natural tendency will be to abandon restraint, expand its definition of vital interests, and inflate its assessment of its capabilities. The United States could react in such a way as to reinforce Russian views of the United States as its main adversary. A US policy agenda devoid of human rights and economic liberalization would also experience tough going domestically. It would be all too easy to find ourselves again in a highly conflictual, zero-sum relationship.

These scenarios are designed as planning tools, to help us think systematically about US policy should the future begin to track this scenario. This does not imply that the United States has decisive leverage over Russia's future, and indeed in this scenario our leverage is modest. But we may be able to tilt the direction of change in more favorable directions, if we can accept the plausibility and significance of "working authoritarianism."

Broadly, two policy responses are possible should the scenario begin to unfold. One would be to resist the emergence of a successful Russia, using whatever leverage we had over its strategic partners, and accepting whatever conflicts resulted as less costly than the reemergence of an aggrieved, powerful peer competitor. However, Russia's success in this scenario depends primarily on its own internal actions, and on strategic relationships with countries that are acting with increasing independence in, for them, a less threatening environment. Preventing such an outcome would absorb enormous diplomatic capital, have minimal effect on Russia's new partners, and guarantee Russian hostility.

The preferable approach, and the one in keeping with the current administration, is closer to "limited partnership"—a pragmatic, issue-specific, weakly institutionalized set of discussions and negotiations, the net effect of which could be a more cooperative relationship. The benefits of this relationship to both parties would hopefully be sufficient to help contain the damage from the inevitable conflicts—in resources, trade, the Middle East, and so on—between the United States and a successful Russia.

The "Strong State" scenario reproduced below is the China equivalent of the Russia scenario. It describes a mostly successful economic transition from an investment- and export-driven growth model to greater reliance on domestic consumption, producing lower but still adequate levels of growth. Political reforms are limited but effective and, in combination with a crackdown on corruption and growing middle-class prosperity, maintain the Communist Party's political legitimacy. The scenario fits real events to 2015 quite closely, and poses a more credible threat to the Western model of democratic capitalism than does the Russian scenario.

---

## CHINA: STRONG STATE

The Chinese system of government remains authoritarian, with some elements of controlled democratization.[28] It has met its economic commitments to the Chinese people by sustaining robust growth while slowly rebalancing the sources of growth away from investment and exports. This remains a key priority.

The Chinese Communist Party (CCP) continues to govern as if it were part of a contested party system, although there is no traditional opposition. Instead, it faces resistance and feedback from an array of challengers including Hong Kong, Taiwan, the entrepreneurial classes, the rural masses, and external actors—all with expectations and demands that force the CCP to continuously review its policies and adjust when necessary. With its focus on survival, the CCP closely monitors the sources of its legitimacy, using polls to determine if overall CCP policies are generally supported and also to track public opinion about the performance of senior party figures. How a particular provincial leader is polling will determine how he or she progresses within the party. Collective leadership and the party's dominance are helped by the long grooming process and by the principle that nominations to the top are always vetted by the generation before last—effectively denying complete control to any particular group of leaders. Such feedback enhances the quality of governance, enabling leadership to competently manage the challenges of inequality, corruption, and environmental blight. When necessary, its enhanced capacity allows it to successfully suppress dissent and maintain law and order.

The CCP uses a top-down approach in managing the dissemination of information to the general public. Following any incident such as a natural disaster, a food safety scandal or the uncovering of corruption, the CCP manipulates public opinion by allowing people to vent and express their anger for a short period of time. This creates the illusion that freedom of speech is tolerated. However, media coverage is always closed down within a few weeks, thereby preventing matters from festering. The public barely notices the media blackout and is left feeling that its voices have been heard.

The modernization of China is at the top of the CCP agenda in 2020. The government has embraced many new technologies during the previous ten years, and technology has become a key government enabler. The CCP has issued electronic ID cards to the public and moved to a predominantly e-government environment. State security makes full use of Internet and mobile tracking, while China's video surveillance system has become the world's most sophisticated.

This has improved the quality of government reporting, leading to better decision-making. It has introduced a much higher degree of transparency to public interactions with government officials, thereby making requests for bribes and other forms of corruption more difficult to hide. While improving efficiencies in the planning and delivery of public services, the ID cards are also a very effective government tracking tool and are used to follow the movements of the vast Chinese population.

Critical to safeguarding party legitimacy has been the CCP's successful campaign against corruption. While China still does not have an independent judiciary, the party has implemented structures to encourage the reporting of corruption and ensure swift and effective justice for the perpetrators. Its evolving bargain with the Chinese people remains intact.

## Drivers of this Scenario

The strong state scenario is made possible by the interactions of the following drivers:

### Economic Trends

The CCP delivers sustainable economic growth of between 8 and 9 percent over the ten-year period. It adjusts its interest rate and currency policies to gradually move the economy away from its heavy dependence on investment and exports toward increased domestic consumption. While the population no longer sees the double-digit growth of the 1990s and 2000s, it accepts that the new policies implemented by the CCP are necessary for the longer-term health of the Chinese economy.

### Corruption

The CCP introduces a new internal department reporting directly to the premier, which has as its sole objective the elimination of all forms of corruption. While it is viewed as a long-term program, the party proclaims a zero-tolerance policy and engages the public directly in exposing those guilty of corruption and ensuring that they are brought to justice. Selective enforcement of this policy enhances the insecurity of citizens and encourages political conformity.

*Inequality*

By delivering sustainable economic growth, the CCP provides a degree of stability to the Chinese population. The additional focus on the development of Western China brings new opportunities to those populations and closes the income gap between them and their counterparts in the coastal cities. It eases tensions within the population.

*Information Technology*

Technology becomes an important government enabler, allowing it to improve the quality of decision-making and eliminate many inefficiencies and waste from government departments. It also becomes an important tool for the CCP in suppressing and co-opting opposition.

*Civil Society*

Tensions grow between the CCP and civil society primarily due to the CCP's continued censorship of all media outlets in China and its tight control over Internet usage. NGOs remain highly regulated but are allowed to fill gaps where their knowledge and expertise is of value to the CCP and can enhance overall state capacity.

*Demographics*

The introduction of electronic ID cards across the population enables the CCP to track demographic trends more accurately and start to respond more appropriately to public requirements. The addition of highly trained demographers to senior positions within the CCP ensures that new policies are introduced to try to prevent worst-case scenarios from unfolding.

*Environment*

Energy conservation and environmental protection become central to all aspects of China's economic activities and key motivators in the movement of economic activity away from heavy industry.

### Changes within the CCP

The CCP manages to refresh itself. It doesn't succumb to corruption or expectations that it will lose power. Instead, it reforms organically from within. It encourages the best and the brightest to step forward, which, in improving the quality of government, reinforces CCP survival.

### International Economy

China has used its terms of entry into the WTO in 2001 to build up huge, internationalized Chinese state enterprises that challenge established multinational companies and act in tandem with the government. This rewrites the story of globalization with an economic power shift toward China.

## The Path to 2020

### 2010: A Start Is Made to Rebalance the Economy and Address Inequality

The 2008 stimulus package was an overwhelming success, when viewed in the context of its primary objective—to reignite economic growth. However, the CCP was aware of the potentially damaging consequences of its heavy focus on investment, and in 2010 it started to take corrective action. The stimulus program had kick-started the economy by investing heavily in large infrastructure projects, which had the immediate impact of increasing demand for heavy industrial goods and creating new jobs. However, much of the funding was inefficiently used and led to the creation of overcapacity in many industrial manufacturing sectors. By mid-2010, production significantly exceeded domestic demand, and much of the surplus production found its way into European and other developed country markets, often at substantially discounted rates. This development led to an escalation of official complaints against dumping and trade actions against Chinese products. In parallel, the loan component to the stimulus had created dangerously overinflated property and equity markets. The CCP acknowledged that these markets needed to be cooled down, that capital needed to be directed more efficiently to take advantage of China's pool of surplus labor, and that it needed to introduce new policies that would stimulate sustainable growth in domestic consumption. The CCP also needed to demonstrate progress toward its commitments to reduce the energy intensity of Chinese economic growth. A continued focus on heavy industry would destroy any prospect of reducing carbon emissions.

By late 2010, the CCP had introduced new policies that would be expanded as the decade progressed.

- New property taxes were introduced that made
  speculative investment in property less attractive.
  This had the immediate effect of slowing down the
  construction frenzy, particularly in the major coastal
  cities. New capital gains taxes were applied to stock
  market gains, which helped to cool down equity markets.
- The CCP initiated a process of slowly increasing interest
  rates. This had the effect of making the capital previously
  used for investments in heavy infrastructure much
  more expensive. As a result, entrepreneurs diverted
  more funding into less capital-intensive and less
  carbon-intensive service industries, which in turn created
  more jobs.
- Higher interest rates led to greater consumer confidence.
  The Chinese people saw their savings accounts earn

interest, which made them feel more financially secure and therefore less conservative about spending.

- The year 2010 also marked the start of a change in currency strategy. While never publicly conceding to US and EU demands, the CCP began to gradually revalue the renminbi. This helped the CCP economic policy on a number of fronts. By making the currency more expensive, it increased the cost of Chinese exports and decreased the cost of its imports. This helped reduce the economy's dependence on exports as a source of growth and gave Chinese consumers access to new imported products, driving domestic consumer demand. This change in strategy also lessened tensions with international trading partners, as it helped to reduce trade imbalances.

- Aided by new international standards for the reduction of emissions and international enticement schemes to developing countries, China's industrial policy aggressively pushed alternative energy and environmental technologies, including on the export front. Made-in-China solar panels, wind farms, and electric batteries started to capture the next wave of consumer technologies.

In 2010 the CCP was presented with a new domestic challenge, when a terrorist attack at the World Trade Fair Expo in Shanghai killed over one hundred people. The CCP responded with a determination that left no stone unturned. The perpetrators from the Xinjiang region were captured and executed. The CCP increased its surveillance activities over suspected members of the Xinjiang and Tibetan separatist movements. It launched an aggressive publicity campaign to remind the Chinese public that terrorism now posed a major threat to

Chinese security. The "Safe Streets" initiative, initially launched in the winter of 2006, received increased government funding to dramatically expand the web of surveillance cameras operating nationwide, particularly in Xinjiang and Tibet.[29] Originally developed to detect possible illegal demonstrations, the program's mandate was expanded to include monitoring for potential terrorist activity. However, the CCP recognized that surveillance was not enough and that it needed to address the issue at its source.

It understood that poverty provided a ripe recruitment ground for terrorist organizations. The tenth anniversary of the Western Development Strategy had been celebrated in January 2010. Following the terrorist attack some months later, the CCP recognized that following through on the proposals outlined by Wen Jiabao at the Western China International Cooperation Forum in October 2009 was now critical to undermining all separatist movements. Premier Wen had outlined his proposal for the further development of Western China under four headings:[30]

The deepening of energy cooperation and transportation links between Western China and its neighboring countries, which were similarly rich in natural resources

The deepening of economic ties through increased cooperation in trade and investment, including a relaxation in import quotas and the creation of new industrial hubs

The deepening of cooperation in energy conservation and environmental protection. The CCP engaged a range of NGOs as consultants and to implement new programs in this area

The deepening of regional and international cooperation through organizations such as ASEAN and the Shanghai Cooperation Organization (SCO)

From 2010 onward, the CCP started to follow through on its commitments, and slowly the western provinces started to see increased levels of investment and the emergence of new local employment opportunities. Although the income gap with the coastal provinces remained, by 2020 it was not as pronounced as in the past, and the CCP carefully manipulated public opinion to ensure a general perception of progress and continued loyalty to the center. Over this same time period, although the separatist movements struggled to build momentum, the CCP kept a close eye on their activities, ensuring that any semblance of agitation was quickly quashed.

## 2011: Fighting Corruption

In early 2011, a dangerous leak from a nuclear plant in Guangdong province led to a major change of strategy within the CCP. Unlike previous incidents in which public safety had been compromised, the CCP did not simply bring those responsible to justice. It initiated a major internal restructuring that was to transform how the CCP would deal with corruption going forward. The CCP recognized that corruption posed one of the biggest threats to China's economic growth, to the safety of the Chinese public, and to the survival of the CCP.

- The premier announced the creation of a new anticorruption department, which reported directly into him. This well-funded new agency was ruthless in exposing fraudulent practices and in ensuring that the maximum penalties possible were secured for those convicted.
- The CCP made promotion up its ranks dependent on a proven track record of fighting corruption. This slowly helped to change the party culture by creating a

degree of competition among party members to expose wrongdoings.

- The use of polling among the public became an important way of monitoring how committed party members were to wiping out corruption.

- The government accelerated the migration of its day-to-day paper-based administrative activities to an e-government environment in order to increase transparency.

- Electronic ID cards were rolled out nationally to simplify interactions between members of the public and government agencies. It made the payment or receipt of bribes by public officials much more difficult to hide.

- The public was encouraged to report instances of corruption via a centralized government website.

- Independent NGOs were invited to work with government departments and state-owned enterprises (SOEs) to introduce new quality and safety procedures that would qualify for international accreditation from the International Organization for Standardization (ISO).[31] It was hoped that clearly defined, electronically trackable processes would act as a deterrent to corrupt practices and that ISO accreditation would improve the image of Chinese-made products.

Given how entrenched corruption was within the Chinese system, it was accepted that progress would be slow and that success would be elusive. Nevertheless, this aggressive new policy was viewed by the public as evidence of the CCP's commitment to addressing this major problem. The anticorruption framework announced in 2011, was fully operational just two years later in 2013.

## 2012: A Leadership Transition

The year 2012 was critical for the CCP with the new generation of Chinese leaders coming to power and presidential elections taking place in both Taiwan and the United States. While sensitive to its own vulnerabilities during this transition of power, the CCP was also very alert to activities in both Taiwan and the United States. When rumors started to emerge that Ma Ying-jeou might not seek reelection in Taiwan, the Chinese intervened immediately to persuade him to run for another term. Although the pro-China movement seemed to have gained considerable ground under Ma's leadership, it was still in its infancy. The CCP needed President Ma at the helm for at least another term, to reinforce his policies toward the mainland.

Meanwhile in the United States, the Obama administration was facing a difficult re-election campaign, with unemployment running at over 11 percent and a growing populism. The CCP was concerned that trade relations could be jeopardized by the American union movement and elements of the US media. In response, Hu Jintao organized a state visit to the United States, using this trip to remind the US administration and the US public of the important partnership between the United States and China. He reiterated the key steps China had taken during the preceding two years to promote a return to global economic stability and to open up its domestic markets to imports from the United States. He outlined the major changes that had been introduced to eliminate corrupt practices and improve the quality and safety standards employed by Chinese manufacturers. He reaffirmed China's readiness to play a military role against threats to sovereign states, particularly in Africa and South America. He spoke of China's support of a strong Pakistan state and its commitment to ongoing development in Afghanistan.

But as well as these positive gestures, China was able to leverage its hard currency war chest: trillions of dollars, partially hedged by increased holdings in raw material and energy resources. The world, and singularly the United States, had come to accept a partial internationalization of the renminbi, based on its use within the Asian region and with developing economies, without China opening itself up to the risk of a fully convertible currency. At the UN and on key international issues, China increasingly held sway over a growing group of countries that often resisted American policy preferences. It had gained a kind of parity, not in hard power, but in the management of the world economic system and in addressing global issues of high importance to the United States. The main theme of President Hu's visit was that the United States and China needed each other to remain strong. Their economies and global interests were so intertwined that if one were to fail, both would fail.

The visit was deemed an overall success. Obama committed to continue to work together with China to address many of their shared global challenges and subsequently went on to be re-elected. Meanwhile, Hu Jintao and Wen Jiabao were succeeded by Xi Jinping and Li Keqiang as the new president and premier of China.

Power transitioned smoothly to the new Chinese leadership in 2012. The new leaders continued to drive the previous administration's efforts to rebalance the Chinese economy, fight corruption, and narrow the income gap between rural China and the coastal cities. Top-down engineering continued to be the CCP's standard approach, and from 2012 onward a renewed focus was placed on bringing highly qualified and capable people to positions of influence and leadership. The CCP made sure that high performers in the provinces, in central government departments, and in state-owned enterprises were recognized and rewarded. Exceptional performers were promoted. This renewed focus on developing and promoting talent within the CCP was crucial to improving the quality of Chinese policymaking. It also improved intraparty cohesiveness and loyalty. The CCP had its best economists steer the country's interest rate policy, drive its currency revaluation policies, and carefully manage economic rebalancing. It engaged the country's brightest scientists to drive its policies for dealing with climate change. It engaged the country's most knowledgeable demographers to drive the required changes in fertility policy and in the development of a state pension program to help support the rapidly aging population.

The Chinese state apparatus had brought China to its current position as a global economic powerhouse. However, there were huge inefficiencies that could now hamper China's progress to the next stage of its economic development. A key challenge for the CCP was how to combine its surplus capital with its surplus labor. It would need a strong reformed financial system underpinning these efforts. These objectives became central to CCP strategy for the remainder of the decade.

### 2013–2020: The Role of Technology and an Evolving Relationship with the Public

Technology was embraced by the CCP as a critical enabler in accelerating the achievement of its ambitious targets and in enhancing state capacity.

Between 2013 and 2016, the CCP rolled out its nationwide electronic identity card program. This proved highly successful. It gave the central government full visibility of the entire population. It provided important information on healthcare and education needs, on demographic trends and demand for public services. It improved the quality of government reporting and city planning activities. It enabled the government to keep a closer eye on the vast population, and improved its responsiveness to the population's evolving needs, which further enhanced CCP legitimacy. As government departments moved to an electronic system of management, waste was reduced and transparency increased. The move to e-government transformed government procurement activities and eliminated many of the corrupt practices of the past.

Technology also enabled challengers to the CCP. To combat dissident groups, the CCP spent vast amounts of money on enhanced Internet security and surveillance equipment. It continued to invest heavily in new technologies to monitor Internet usage, detect dissident activity, and close down threatening sites. It infiltrated many of the underground networks by using the same technologies as the bloggers and activists themselves. The CCP also expanded its use of surveillance equipment nationwide.

This market grew from an estimated $43.1 billion in 2010[32] to $500 billion by 2020. From its origins as a tool in the fight against crime and an early warning system against possible public demonstrations, surveillance cameras in public places were increasingly used to track targeted groups of individuals. Improvements in face recognition technology made it possible for the police to track an individual's movements throughout an entire day.

The CCP's heavy-handed, top-down approach created significant tensions with civil society. The government continued to strictly censor all forms of public media and Internet usage. It was accused of blatant human rights abuses against individuals who dared to challenge its rule. It maintained tight control over the registration of nongovernmental organizations and was ruthless in closing down those that challenged CCP policy. However, over the decade the CCP was astute enough to recognize that in certain niche areas, state capacity would be enhanced by engaging NGOs to perform certain tasks. NGOs supplemented state-provided health and education services, helped the CCP to improve the processes used by government agencies, and helped a number of SOEs secure international ISO accreditation. NGOs were also successfully used by the CCP in energy conservation and environmental protection, supplementing increased levels of cooperation with the EU and United States on energy technology development and capacity building.

As the decade progressed, China was emerging as a clear leader in the global clean energy technology race. Leading the world in clean vehicles, solar and wind technologies, and the development of game-changing carbon capture and sequestration technologies enabled China to grow new export markets and create new domestic jobs. This further solidified the impression of competence as the central government executed a highly effective long-term energy strategy. A critical driver of the clean energy revolution in China was the adoption of a carbon tax starting in 2013.

## Public Opinion

From 2015 onward, the CCP expanded its use of polling to keep its finger on the pulse of the population. New technology and an increasingly IT-literate population made electronic polls a highly effective tool in tracking public perception of government and individual party members' performance. Polls were used to determine what the public thought about local leadership, the campaign against corruption, and what areas were being neglected by the government and needed more attention. While the public did not directly elect party members to government, citizens increasingly saw that their opinions were having an impact. Provincial leaders who continued to engage in corrupt practices and who were not performing satisfactorily lost their jobs. Those leaders perceived by the public to be doing a good job tended to be promoted.

Over this period, there was a growing sense of social justice as corrupt politicians were caught and convicted, investment and growth opportunities were now being directed to western provinces and income gaps were starting to narrow. While far from the Western ideal of a true democracy, the Chinese system was viewed by the population as increasingly participatory. Their voices were being heard. Within the CCP, there was also a feeling that things were changing. While intraparty democracy was still in its infancy, a culture of meritocracy was gradually filtering down the organization. High performance was starting to replace the old culture of cliques and personal connections as a determining factor in appointments, leading to much greater transparency in the workings of the CCP.

## Conclusion: Implications for US Policy

While China remains highly autocratic in 2020 with the CCP firmly in control, the Chinese public is still generally supportive of the central government. The Chinese leadership has brought in highly qualified and effective people to address the many significant economic, demographic and environmental challenges faced by China, and they seem to be making progress. The internal structure of the CCP has evolved to deal with corruption. The CCP has embraced technology as a key enabler of strong government. It has become a more constructive stakeholder in the international system, albeit a system over which it exerts growing influence, committed to and capable of discharging its international commitments.

The consequences for US policy are, as in other scenarios, both positive and negative. On the positive side is a China committed to reducing its dependence on export-led growth, thus potentially a more effective partner in managing the global economy, generating growth and poverty alleviation throughout the global system, and maintaining a more balanced trade and financial relationship with the United States. Also, the government has the domestic writ—though not necessarily the motivation—to deliver on any commitments to trade and currency liberalization, reduced carbon emissions, nuclear nonproliferation, and regional stability.

Realizing this upside will, however, be very difficult. On the part of the United States, the urge to see and encourage economic and political liberalization in China will interfere with even "à la carte" cooperation with a strengthened Chinese state. For China, with a legitimate and robust government, a GNP approaching the United States', and enormous financial resources, the obvious tendency will be to rediscover dormant ambitions and to pay lip service to reform of the Western order while quietly (and eventually not so quietly) seeking to overthrow it. Asian regionalism in trade and monetary affairs will deepen. The contest between democratic and state capitalism could echo the Cold War in its reach and intensity.

This suggests a complex, rapidly changing relationship, full of ambiguity and uncertainty. Objectives on which the two states agree—counterterrorism, stabilization of failing states, promotion of global growth, avoiding worst-case scenarios for climate change—might be greatly advanced by cooperation. But the United States' partner at times will appear as a rival in setting the rules for the global economy, building a strong nonproliferation regime (as opposed to ad hoc cooperation with regard to certain states), and maintaining regional stability (not when instability serves expanding Chinese interests, or when it can free ride on US stability efforts). As Chinese power continues to grow and its state strengthens, finding the right balance between engagement and containment will be a constant challenge for the United States.

## Making Decisions in This Most Likely Future

Decision-making in such an environment is about as hard as it gets. Jeffrey Legro, who has written extensively on surprise and its effects on grand strategy, notes that "an increasingly complex and dynamic world makes designing foreign policy a difficult task. But the same complexity also makes the need for understanding and charting possible futures more important. Government officials with pressing daily demands can rarely look too far ahead; still, as George Kennan illustrated in his logic of containment, a long-term view can be essential to successful strategy."[33]

To decision-makers overwhelmed by the volume and velocity of information, selectivity becomes a necessary coping mechanism. Ambiguity and information overload force us to be restrictive about what we credit as relevant trends and events, what we choose to look for and observe; and we select based on often unrecognized or unarticulated assumptions about the future.[34] These assumptions derive from recent experience (which can produce misleading historical analogies or trend extrapolations), a commitment to current policy and the assumptions necessary to validate those policies, time pressure (rewarding assumptions that are "good enough" to permit closure of debate), mindsets about how the world works (based on theory, cultural bias), groupthink (and the political risks of dissent), and the political demands of building a case for change (which create strong incentives to extract the most possible value out of current policy). Foreign policy debates proceed within a context of insecurity, which can encourage threat inflation and actions that produce self-fulfilling downside prophecies. US policymakers are particularly susceptible to these tendencies, given multiple US interests (and the consequent thinning of intelligence and increased uncertainty), and the magnitude of relative US power in the world, which multiplies perceived threats, can blind us to the interests and perspectives of others, and, when deployed carelessly, can produce massive unintended consequences. Foreign policy is thought about and implemented through stovepiped organizations, often unable to discern the interaction of drivers, where big changes can occur (globalization, volatility, diminished growth, fragile government), and how these forces intersect in specific countries.

Major, unanticipated change is often a product of trends and events from different domains of activity intersecting in unexpected ways. Indeed, a theme running through expert commentary during the post–Cold War era is an underestimation of the potential for radical change in the complex internal politics of seemingly stable states, with revolutionary consequences for their external capabilities and strategies. Authoritarian states appear impervious to change, until suddenly they do. Expectations of stability, in retrospect, turn out to have reflected limited information, embedded mindsets, and excessive caution. This does not constitute a general prediction of imminent instability, but recognizes that states are today subject to an extraordinary combination of internal and external demands, a consequence of increasing globalization and, recently, of extended global economic crisis. These stresses can unravel governments, both democratic and not, with limited popular legitimacy and high exposure to global volatility.

Ideas, theories, hypotheses, historical analogies—all are essential intellectual equipment for making sense of the stream of events. Policymakers cannot be agnostic about the way the world works. They must look for signals in the noise, try to explain causation, discern patterns, draw conclusions, and make decisions—all in a time frame imposed by events and at odds with the deliberate processes of data gathering and hypothesis testing of academic and intelligence analysts. However, mindsets appropriate for the world as is can themselves become sources of surprise in the presence of rapid change, and thus must be continuously re-examined in light of new data. The problem of policymaking in uncertainty is not lack of a futures perspective among policymakers, but the unexamined, often unarticulated assumptions behind policy, and their stubborn resistance to change in light of negative feedback from the world. They are too easily overtaken by fast-moving events, yet persist into obsolescence.[35]

## Can We Do Better?

Few of these sources of distortion and policy error are "fixable"; making fateful decisions in uncertainty guarantees mistakes, some of them consequential. What we can reasonably hope to accomplish is to make assumptions more explicit, reduce surprise, imagine and mitigate risk, rehearse and thus improve policy responses to wild card events, and recover rapidly from—even take advantage of—the blindsides.

One could imagine several strategies for reducing surprise. American grand strategy could try to do less, thus excluding some of this complexity from our definition of what matters to US interests (another way to reduce uncertainty, "creating our own reality," revealed its hazards in the aftermath of our invasion of Iraq). We could, for example, choose to view liberal globalization as growth maximizing, and therefore politically self-sustaining (in the face of clear evidence to the contrary); much of the developing world as too ill-governed, resistant to our influence, or peripheral to core interests, to be worth our attention; issues of the global commons as amenable to market-based solutions; rising powers as more threatening to—and thus contained by—immediate neighbors, thus less threatening to us; the risk of terrorism as low; and the cost effectiveness of improved homeland security plus covert strikes vastly greater than regime change and state building. Such grand strategies, variously labeled as "light footprint," "restraint," "offshore balancing," "selective engagement," "retrenchment," would, over the short term, reduce the knowledge requirements of US policy. They would increase the likelihood that US actions in the world would have somewhat more predictable effects, and enable intelligence to worry only about "known unknowns" and more accessible and familiar targets.

Doing what we have the knowledge to do well can minimize short-term risk and is an important criterion to be weighed in making policy choices. But as I argued in chapter 1, anticipating and managing risk is essential to sustaining a strategy of restraint. Confining our attention to the better-known world would over the longer term expose the country to great harm, inviting adversaries to fill the gaps left by America's retrenchment and guaranteeing unpleasant surprises from outside the restrictively defined perimeter of our interests. This approach, rejected by administrations from both parties over seventy years of Cold War and post–Cold War history, has now found an advocate in President Obama ("Do no harm," "Lead from behind," "Don't do stupid stuff," etc.), but the returns on this investment—including mounting threats in Iraq and Ukraine—suggest the great difficulties of managing a process of retrenchment without accelerating the erosion of our influence and losing domestic support. It is difficult to imagine the next administration, regardless of party, embracing such an approach. Warranted or not, the temptation to engage actively in the global system, and thus to encounter many of the sources of uncertainty, complexity, surprise, and risk discussed above, appears irresistible to a country that is powerful, vulnerable, and idealistic. Getting better at decision-making in these contexts, as we deepen our knowledge about the world, is an unavoidable necessity.

A more promising approach to uncertainty requires that we take a closer look at the concept of surprise. As I've argued, surprise is inevitable in the world that is emerging. But some events are surprising because we are cognitively unprepared for them, or find them too politically inconvenient to entertain. We call them surprises because we are surprised by them, whether lead indicators were present or not. By failing to differentiate between events that are inherently surprising and those that are more or less foreseeable, we give decision-makers a free pass on myopia or willful ignorance.[36] Indeed, few of the surprises noted earlier were bolts out of the blue, without leading indicators or even a few prescient observers. They were gray, not black, swans. They were the result of the interaction of underlying trends and of incremental change unseen by experts. They were in part self-inflicted, representing a failure to recognize or properly weigh new information. They were consequences of unchallenged mindsets, of an excess of self-confidence operating in a dynamic world. We were, in many cases, blindsided by our own assumptions. Foresight is not the point here; acute observation was lacking. And our subsequent reactions to surprise have often demonstrated more stubbornness than agility, thus greatly magnifying the damage from the inherent limits to foresight. As former president of the New York Fed and US Treasury secretary Timothy Geithner said in describing the Fed's reaction to the financial crisis, "Policy was always behind the curve, always chasing the escalating crisis"[37] Or, from an earlier period in history, Sir Anthony Eden: "We are all marked to some extent by the stamp of our generation[;] mine is that of the assassination in Sarajevo and all that flowed from it. It is impossible to read the record now and not feel that we had a responsibility for always being a lap behind. . . . Always a lap behind, a fatal lap."[38]

Unchallengeable assumptions, mirror imaging (other countries think like us), wishful thinking, entrenched policy positions, bureaucratic inertia, lack of imagination—all have played a part in the "intelligence failures" and policy missteps of the last twenty years, and have been on public display recently in our extemporaneous response to the Arab Spring, and in the "lap behind" performance of the Obama administration in Syria and Iraq, and in Ukraine. These are the types of surprises that improved process can mitigate.

Surprise can be reduced, through imagination, informed and continuous observation, and learning. Nor do surprises—warranted or not—necessarily produce bad policy if decision-makers are prepared to modify outdated assumptions in the presence of unpredicted events, and have prepared cognitively and strategically for a changed environment. Indeed, preparation for surprise can allow leadership to productively exploit the political capital that often becomes available in the aftermath of shocks. Such shocks afford the opportunity to rethink basic assumptions and reformulate grand strategy in a way that incremental change cannot.

## Chapter 2

# Historical Cases

History is replete with examples of governments pursuing strategies premised on flawed assumptions, governments that, in the most extreme cases, disappeared as a result. Hitler assumed that he could knock out Soviet Russia in six months; imperial Japan assumed that the United States would not fight but sue for a negotiated settlement following Pearl Harbor; Gorbachev assumed that political and economic reform of the Soviet Union could be managed and contained. Lest we be tempted to conclude that authoritarian regimes are uniquely susceptible to strategies premised on questionable assumptions, benign European assumptions about Nazi intentions were nearly suicidal. A few concrete examples from America's recent history should make the case for the damage done to our interests by assumptions about the world that were either wrong from the beginning or had outlived their usefulness.

## The Disintegration of Yugoslavia

The Balkan unraveling following the German recognition of Slovenian and Croatian independence was at least a plausible outcome, and indeed predicted by some in the British Foreign Office. The conviction of two American presidents that we had "no dog in that fight" were oblivious to worst-case possibilities of the irredentism and violent ethnic conflict that came to pass, the consequent loss of life, the threat to the viability of NATO, and the need for a costly (and continuing) intervention. These possible outcomes could have been more forcefully argued during the seemingly endless months of equivocation that preceded the decision to forcefully engage. That the Clinton administration succeeded in negotiating a mostly peaceful partitioning of the former Yugoslavia does not vitiate the fact that with a more timely recognition of the downside risks of escalating Balkan conflict, preventive/coercive diplomacy might have avoided much of the carnage, and a large and ongoing commitment. Collateral damage to US-Russian and US-China relations might also have been minimized. That this first post–Cold War conflict was unfamiliar, institutions unprepared, and responses halting and extemporaneous, is precisely the point of alternate futures as a policy tool: to make the unfamiliar and unexpected plausible and consequential, as the basis for thinking about prevention or mitigation.

## The Failure of the Doha Round of Trade Negotiations

International trade negotiations provide another example. The underlying and unarticulated US assumption entering the 1999 Seattle meeting, which was to launch another multilateral round of global trade and financial liberalization, was utterly at variance with the world that was then emerging. The existing policy paradigm called for a US-crafted agenda of extensive liberalization by emerging countries in such areas as financial and other services, investment-related performance requirements, intellectual property, government procurement, subsidies, and antitrust. These new areas of liberalization were to be balanced by wealthy country trade concessions of interest to poor and middle-income countries, specifically in agriculture and labor-intensive manufactured products. The first post–Cold War round was understood as another in a series of multilateral trade negotiations to advance the liberal agenda toward a truly borderless world, stimulate global growth, and further enhance the legitimacy of the US-led, rule-based trading system. The multitude of cross-cutting interests and conflicts was to be mediated by US leadership and leveraged by the implied threat of US market closure.

None of this came to pass. As the Doha Round still gasps for breath, it has become clear that the advanced countries are not prepared to liberalize agriculture. The emerging industrial powerhouses of China, India, and Brazil have balked at liberalization in manufacturing, fearing competition from less-developed, lower-wage locations. The so-called Singapore issues (intellectual property rights, investment distortions, free trade in services, etc.) are nonstarters for many developing and emerging countries, who view these demands as an effort to impose a made-in-America development model and to dominate new sectors of the global economy. US economic diplomacy, exercised previously in a bipolar world of limited GATT membership and strong strategic incentives to collaborate, has found its power greatly diminished among a much larger number of WTO members, with reduced leverage over its own allies (and with less impetus from US business, which saw only modest value added in further liberalization).

An earlier recognition of these emerging trends and their consequences for global trade liberalization might have produced several benefits: a more modest diplomatic investment in American liberal hegemony, more in keeping with diffuse power and competing growth models; a more collaborative style and a more modest agenda in multilateral negotiations; a quicker reaction to Asian regional trade developments, which have now come into focus in our belated embrace of the Trans Pacific Partnership; and a reallocation of our diplomatic capital toward global economic issues of far greater importance for growth and stability than incremental trade liberalization: global financial imbalances, currency misalignments, cross-border regulatory issues, cyberthreats to open trade and investment, and transnational crime.

One would not expect an alternate scenario exercise to have produced a revamping of traditional American trade policy in advance of Seattle. The value proposition is that such an exercise would have provided an alternate context for understanding the growing evidence that the current policy was experiencing diminishing returns, that others' resistance to further trade reform would not succumb to a last-minute US diplomatic blitz. It would have expedited the debate about the costs of doubling down on our commitment to multilateral rounds of trade liberalization and suggested alternate courses of action. It would have enabled policy to be less reactive.

## The Terrorist Attacks of 9/11

The terrorist attacks of September 11, 2001, revealed deep flaws in prevailing assumptions about the degree and source of risk the country faced at that moment, based on the administration's realist assumptions about the primacy of state versus nonstate threats. The misdirection of our attention was greatly compounded by misguided reactions to the event itself that indicated lack of thoughtful preparation. Serious attention to warnings from the outgoing administration about the gathering threat from al-Qaeda would not have permitted a clear prediction of the attack. But they could have heightened alertness to the possibility, caused resources to be shifted toward monitoring of terrorist plots, and led to precautionary measures at vulnerable locations. Such actions might have complicated what was already an elaborate plan to attack New York and Washington, forcing the plotters into riskier, and more visible, behavior. The plot, as carried out, was long in the planning and rich in observables, from flight training and document acquisition to the boarding of commercial aircraft for the final attack. Thus, while Richard Clark's initial warnings were not specific to the events that unfolded, they might have launched a process of closer observation, based on hypothetical but plausible scenarios (of which an attack on the World Trade Center would certainly have been one), which might have prevented the occurrence.[1]

This example illustrates what was said in the introduction and is treated more expansively in the next chapter: the purpose of alternate futures is not to predict, but to prepare, by sharpening observation of a complex world and simulating policy reactions to potential threats (or opportunities). In the case of a devastating potential future event, the purpose of scenarios is to make such threats plausible, when denial is cognitively easier and politically more convenient, so that preventive actions can be taken. That we have subsequently foiled several attacks suggests the feasibility of prevention, even of actions by diffuse and secretive organizations. The question is whether we needed the event itself to generate the attention necessary for prevention, or whether a legitimate, well-conceived scenario process could have provided some of the urgency and direction clearly lacking in the Bush administration. Admittedly, Clark himself is skeptical: "America, alas, seems only to respond well to disasters, to be undistracted by warnings. Our country seems unable to do all that must be done until there has been some awful calamity that validates the importance of the threat" (254).[2] Given the increasing costs associated with reacting after the fact, doing better is an imperative.

## The US Invasion of Iraq

The US invasion of Iraq is a narrative of assumptions that were wrong from the beginning. These mistakes were willful, not the consequence of well-reasoned analysis confronted by genuine surprise. The sectarian violence, governmental incapacity, resistance to occupation, and lack of tangible allied support were well within the expectations of State Department and academic experts,[3] but were dismissed by the policy's advocates as reflecting an overly cautious view of our capacity to prevail in Iraq, and the international legitimacy and support that would follow. Under these circumstances, it is hard to imagine the Bush administration benefiting from any exercise in considering alternate futures, even one that precisely imagined subsequent events as they unfolded. On the other hand, such an exercise might have improved the quality of the debate within and outside government, both before the invasion and as its consequences became apparent.

Our first scenario exercise at the Center for Global Affairs at NYU was done in early 2007 and imagined alternate Iraqi futures after the expected withdrawal of US forces (the scenario construction process applied in this and other studies is described in chapter 4). The outcome demonstrated the still valuable insights to be gained by a facilitated alternate scenario process among experts and foreign policy observers. Indeed, the "Contagion" scenario (scenario 3) is very close to historic and still unfolding events in Iraq, Iran, and Syria; the dictatorship scenario (number 1) still represents the best hope for Iraq's stability; and the "contained mess" (number 2) combines plausibility with a "good enough" outcome, but is now an opportunity lost.

---

### IRAQ SCENARIO 1: NATIONAL UNITY DICTATORSHIP

*Summary*

Confronted with the harsh realities of continuing sectarian rivalry and religious violence in Iraq, and with the American military presence dwindling, attention of local and regional players turns toward recreating an acceptable version of Hussein's rule. Minds concentrate on the imperative of keeping the country's nascent civil war from expanding into regional conflict and in finding, or facilitating the emergence of, strong and unifying central leadership. This goal is shared, more or less, by the United States, which has long given up on democracy building in Iraq.

A "national unity dictator" (NUD) is willing and able to suspend the constitution in order to address the lack of law and order that has led many Iraqis to flee the country or to throw their support behind insurgent groups and sectarian militias. Such a NUD would tap into dormant strains of Iraqi nationalism by resisting all elements compromising Iraqi sovereignty, including Sunni insurgents, al-Qaeda in Iraq (AQI), the American presence, Iranian-supported militias, and Kurdish separatists.

The NUD is not necessarily a secular leader and is likely to be situated between all of the ethnic and religious factions that currently divide Iraq. This leader is an Iraqi nationalist, not an Arab nationalist, and emerges after a substantial withdrawal of US troops from Iraq, sometime between 2010 and 2012. By that point the failure of the Islamists and successive weak administrations in Baghdad will have persuaded large numbers of Iraqis to trade the liberal freedoms they have enjoyed on paper since the fall of Saddam for freedom from fear.

The NUD represents the first leader in post-Saddam Iraq who combines the room to maneuver in forcefully dealing with violent insurgents with an ability to operate among Iraq's internecine ethnic and religious divisions. A leader who is properly situated between these groups, most likely a Shia with good relations across communities, could exploit tensions between tribal leaders and foreign jihadis and attempt to unite the country under the task of flushing out al-Qaeda-like terrorists and restoring stability.

Any US attempt to anoint a friendly leader with this role is likely to backfire. Indeed, his emergence will probably come as a surprise. In civil wars, it is often military—not political—leaders who are able to consolidate power due to their prowess in battle. That inherently unpredictable process as well as the necessity that the NUD not be seen as doing the bidding of the United States means that the United States may have to wait for him to emerge independently, then court his support.

The NUD is valuable to the United States only to the extent that he is able to hold the nation together and keep the chaos currently reigning in Iraq from spilling out into a regional conflagration. A withdrawal of US troops is necessary for the emergence of a NUD, but not sufficient. Another important component of this scenario is the Iraqi army, which must be transformed into a force that is representative of the entire country and strong enough to engineer the most likely path for a NUD to assume power: a coup. Providing an environment in which a NUD can emerge will require the United States to maintain a delicate balance both within Iraq and among the other states in the region.

*Conclusion*

A policy of supporting the emergence of a NUD in Iraq reflects the re-emergence of realist thinking on the part of the United States after an ambitious but failed project to bring democracy to Iraq. The NUD scenario represents a rediscovery of the virtues of stability in the Middle East, a refocusing of American power on the fight against al-Qaeda, and a prolongation—at least—of the timeline for democracy building.

Of the three scenarios developed by the CGA group, this is the only one from which a stable Iraq emerges. That stability is by no means guaranteed in the long term. The relationship with the NUD could deteriorate over time, as it did with Saddam, or he could lose his grip on power.

The larger problem of increased Salafism and the impact the Iraq war has already had on the global terror threat will remain despite the emergence of the NUD, and institution building and working toward a more participatory political system in Iraq and the wider Middle East should remain the long-term goal, with the NUD thought of as a stopgap measure to help that process take place organically.

## IRAQ SCENARIO 2: CONTAINED MESS

*Summary*

The instability in Iraq continues, with the growing confrontation between Sunni- and Shia-governed states fought mostly within Iraq's borders. As Iraq disintegrates into a brutal all-out civil war, neighboring countries, realizing the potential for contagion, go to great lengths to keep the chaos contained within Iraq. Proxy war is the result, along the lines of the Spanish Civil War. However, the chaos in Iraq affords al-Qaeda greater opportunities in the region, and confronting radicalization will be a constant challenge for Arab governments.

In this scenario, all players other than Turkey are prepared to keep the Iraqi cauldron boiling. However, none want it to overflow. Thus, while the insurgency is fed from the outside, there is an unspoken agreement that the conflict will not be allowed to expand beyond Iraq's borders. Such a delicate balance could be tipped if one of the parties within Iraq gains the upper hand. In the interest of keeping the conflict contained, however, there is motivation to prevent this.

Refugee flows, sectarian violence, and ethnic cleansing keep neighboring states involved in the conflict. Since Iran and Turkey are the only real military powers in the region, most Arab neighbors participate through local proxies.

Everyone in the region knows how to muddle through. They've been doing it for the past fifty years, and continue to do so. While Iraq burns, states keep an eye on each other to make sure the status quo ante is not disturbed.

## Conclusion

This scenario is difficult to sustain over time and could potentially lead into one of the others. The challenges are significant: keeping Turkey from retaliating in force against the Kurdistan Workers' Party; keeping Sunni jihadis from taking their fight outside the confines of Iraq; keeping the Sunni and Shia populations from going at each other in neighboring countries while sectarian violence rages in Iraq; and holding the containment coalition together.

Resisting the outflow of jihadist activists and ideas is the most serious problem, even if overt war is contained inside Iraq. Iraq-hardened jihadists will fan out across the region and attempt to undo the stabilizing efforts of regional state actors by radicalizing their populations. They will seek to violently bring down any or all of the Arab regimes as they spread their revolution, and they will enjoy a degree of support among many of the locals. The key to scenario 2 is twofold: regional powers avoid direct state-to-state conflict while repressing internal jihadi activities as well.

Thus, the path toward scenario 1 or scenario 3 unfolds. Should the destabilizing forces emanating from Iraq succeed, region-wide chaos ensues, engulfing the whole Middle East. Alternatively, fearing just such an escalation, all the major players in the region might allow, or even encourage, the emergence of a national unity dictator. If such a figure emerged, it is unlikely that anyone in the region will resist, even if the NUD proves to be less impartial than originally desired. However, given the proven ability of Arab regimes to muddle through under the most trying circumstances, an extended scenario 2 status quo could develop.

## IRAQ SCENARIO 3: CONTAGION

*Summary*

Iraq has descended into outright civil war. Instability spreads throughout the Middle East. The regional players, competing and insecure, fail to cooperate on matters of defense and counterterrorism and prove unable to contain the fighting within Iraq. While US pressure and the limited military capacity of local actors have succeeded in preventing all-out regional conflict between Sunni- and Shia-led states, the proxy war fought on Iraqi territory (scenario 2) spreads to adjoining states through refugee flows, growing radicalization of Arab populations, escalating nonstate terrorism, and the deliberate efforts of regional rivals to destabilize each other's governments.

Existing regimes in the region cling to power, but with insufficient domestic political support or acquiescence to create coalitions and pursue effective balance-of-power strategies necessary to contain the Iraq civil war. Because their appetite for direct state-to-state conflict is limited, many regimes use substate actors to strike at their enemies. Regional rivalries flare up as various players vie for influence and control. Radicalization of Arab populations increases as sectarian strife radiates from Iraq. In these circumstances, unforeseen events—such as an Iranian-style revolution in a major Arab country—could radically alter the political landscape and reorder foreign policy priorities in the region.

Events could easily globalize this regional conflagration. A serious disruption to the oil supply, as the result of an attack on an important oil installation in the Gulf, is a likelihood in this scenario. Such an attack could come in various guises. Terrorists might target the energy infrastructure, with the United States retaliating against Iran as a target. The United States or Israel could also react to any number of Iranian provocations, including its imminent (by 2010) development of nuclear weapons, leading toward a major war.

*Conclusion*

This scenario is not in the long-term interests of any state actor—regional or global—and this fact argues for the higher probability of a stabilizing Iraqi dictator or sufficient regional collaboration to contain the conflict. The more this scenario is accepted as plausible, the greater the efforts of states to prevent it.

What keeps it in play, however, are several powerful forces. Terrorist groups are thriving in Iraq, have regional ambitions, and view chaos as an ally. The Sunni-Shia divide deepens in Iraq and is spreading regionally. Iran has a hegemonic past, similar hopes for its future, and ties to terror groups with strengthening positions in the region. Arab regimes are insecure and often unpopular, face radicalized populations, and are rife with inter-Arab rivalries that complicate balancing against jihdadist or Iranian threats. In 2010, American popularity among Arab publics is at an all-time low, its physical presence is diminished, and its credibility among the region's states—both allies and rivals—is deeply wounded. Under these conditions, it would be imprudent to believe too confidently in the logic of "self-interest" in the Middle East.

## The Arab Spring

The Arab Spring caught most experts and policymakers by surprise, and the consequent lack of preparation is another example of the "lap behind" phenomenon: policy chasing a rapidly spreading, bottom-up revolt with large but ambiguous consequences for US interests. The lack of anticipation, and the scrambling to catch up with events that followed, have prompted some soul searching among Middle East experts. Gregory Gause, a widely respected academic specialist on the Arab world, observed in *Foreign Affairs* that "the vast majority of academic specialists on the Arab world were as surprised as everyone else by the upheavals that toppled two Arab leaders last summer and now threaten several others."[4] His explanation, broadly, is that the long period of authoritarian rule focused experts' attention on explaining the persistence of these regimes, rather than on their vulnerability, and generated policy advice to bet on their continued survival. In emphasizing the strength of the military-security complex and state control over the economy, they missed the growing professionalism of some Arab armies (in Egypt and Tunisia) and the rising influence of a new business class benefiting from economic reforms and globalization, and less dependent on regime patronage. As a result, experts "underestimated the popular revulsion to the corruption and crony privatization that accompanied the reforms."[5] They also missed the power of cross-border Arab identity, thus failing to anticipate the contagion that was a remarkable attribute of the revolution in the Arab world. Gause concludes that while explaining stability was an important task, "it led some of us to underestimate the forces for change that were bubbling below, and at times above, the surface of Arab politics."[6] He calls not for precise predictions but for greater humility, for a thorough and open re-examination of assumptions on key drivers of Arab politics: the role of the military, the effects of economic change, the importance of Arab identity. And we should not presume to control these forces for change, because they originate "in indigenous economic, political and social factors whose dynamics were extremely hard to forecast."

Gause's self-criticism is refreshing, useful, and rare. But beyond the misestimation of particular shifts in the domestic politics of Egypt is a broader question of why experts working over many years in an area so important to American security should have organized their thinking around regime stability and, if this is a tendency associated with country experts generally, how we can reduce the chances of being blindsided by expectations of stability suddenly proven wrong by events on the ground. As I've argued in chapter 1, the combination of continued but illiberal globalization, and diminished US leverage in shaping the politics of key regions, will continuously stress-test governments and states, especially those with deep ethnic, sectarian, or class differences. While this is not a general prediction of instability applied to all such states in all circumstances, it should make us far more alert to the potential dissolution "bubbling below the surface." Stability, or the oft-invoked "muddling through" scenario, should not in the emerging world be the default assumption. Although Arab dictators have made a comeback from the Arab Spring revolts, particularly in Syria and Egypt, they will not enjoy the prolonged stability they once did. In this sense, a US policy bet on authoritarian resilience in the Middle East faces lengthening odds.

Taleb and Blyth suggest that the events in Egypt are a classic example of a "black swan," an event (or series of events) both inevitable and unpredictable: the former, because systems operating in dynamic environments that prohibit incremental adjustments to change guarantee themselves some form of shock; the latter, because such shocks can be precipitated by an infinite variety of events, some seemingly trivial.[7] Since these conditions are not confined to Egypt or to the Middle East, the suggestion is to expect and prepare for the unexpected, not by predicting the inherently unpredictable, but by imagining such events (within the bounds of plausibility), then looking for evidence that they are impending.

The emphasis on explaining continuity, referred to by Gause, raises another possibility, namely a value preference for stability among Middle East policy-oriented experts. The reasons are obvious: a well-justified fear of the consequences of revolutionary change, with the resulting opportunities for Islamists and for Iran; and a level of comfort with Arab dictators amenable to US influence, prepared to cut deals with Israel, and able (seemingly) to maintain effective control over their territory. These policy preferences shape the incentive structure for experts eager to influence the policy debate, and establish a range of acceptable analysis that can brilliantly argue the case for continued stability but usually misses the tipping points. It's always safer to extrapolate, and most of the time more accurate as well. Imagining discontinuous change is intellectually challenging, and can be professionally risky. Better to be correct most of the time, and when spectacularly wrong, to have lots of company.

There is also a disconnect between assessments of stability at the individual country level—and the expertise country experts bring to the table—and the operations of the global political economy. The world faces an enlarged supply of economic and political stress resulting from globalization and its mismanagement, and regimes unable to adjust to these stresses, even those of long-standing, are especially vulnerable to sudden political change. With all the growth benefits of globalization, we often ignore its disruptive effects and have not connected these effects to individual states already suffering from homegrown problems, including—in the case of much of the Middle East—sectarian conflicts, youth unemployment, income inequality, government corruption, and concentration of political power. Volatility, economic insecurity, and rapid shifts in competitiveness are conditions of the market-driven globalization that all states are subject to. Badly governed states can often ride out these storms but are all the more vulnerable when the global economy suffers from extended weakness. The point is that expectations of upheaval are not as far-fetched as long-running stability and global prosperity make them appear, an observation that applies to the Middle East as well as to the latest list of global winners, the so-called BRICS (Brazil, Russia, India, China, and South Africa), an acronym now virtually out of print.

## Obama's Grand Strategy

The Obama administration's grand strategy has embodied several mutually dependent objectives, each premised on assumptions about US power and the emerging nature of international politics.

Great power cooperation, the first key element, was viewed as essential both to discourage great power conflict and to manage shared global problems: WMD proliferation, region-destabilizing state failure, global economic volatility, climate change, and destructive competition for resources. It was expected that pragmatic, cooperative problem-solving and the imperatives of globalization would create a stronger logic than great power competition, despite the pivot and Chinese perceptions of encirclement, or Russia's end of Cold War grievances.

Such cooperation was essential to a second key element in the strategy, restraint in the use of force. Underlying this commitment to restraint were the Iraq and Afghanistan experiences, a fear that even limited military intervention in turbulent countries might lead to a repeat of Iraq, and a perception of limited domestic political and financial capital for carrying out such projects. It was hoped that great power cooperation, if not sufficient to do "nation building," would at least enable joint containment of internal violence in fragmenting states.

The third element of the strategy was a heightened focus on Asia, reflecting a judgment, fully supported by recent trends, that the center of economic activity and political/military capacity had shifted toward Asia, that China was both an essential collaborator and our emerging global competitor, and that American attention and resources, having been overinvested in Iraq and Afghanistan, and now under stringent budget pressures, had been slow to adjust to these new realities. The rebalancing (née pivot) was but a formal and the most recent acknowledgment of a shift in American policy that had been underway, haltingly and incoherently, for several years, encompassing military deployments, regional diplomacy, the international legal regime, and trade/investment strategy.

The judgments underlying this three-pronged grand strategy may have been reasonable hypotheses at the time, plausible operating assumptions based on evidence and on the administration's convictions that were necessary to proceed with a foreign policy. Global problems do require global solutions, if they are to be solved or managed. Great powers do have a common interest in addressing them, and rational leadership—as we define the term—should be amenable to persuasion on this agenda. The invasion of Iraq was a mistake of historic proportions, with consequences still reverberating, and not to be repeated. America has lost its appetite for foreign intervention and is now ready for internal rebuilding. There does remain a powerful argument that Asia deserves greater US attention and resources.

The issue is how to conceive and implement a strategy to this end, which is honest about—and therefore better able to manage—downside risks, avoids damage to other vital interests, and makes explicit and plausible assumptions about the "drivers" that enable/require the strategy and could, as they shift, demand its revision. In a period of rapid change and wide uncertainty, even the most coherent of strategies can be premised on assumptions that are subject to quick obsolescence. Yet once such strategies are adopted, they acquire political/bureaucratic legitimacy that can extend their life well beyond their value, distorting perceptions of the world and narrowing the range of politically acceptable debate even as the strategy produces diminishing benefits or perverse, but unobserved, consequences.

Among the various assumptions underlying Obama's grand strategy, the expectations for great power collaboration have turned out to be the most unrealistic, and the administration slow to disabuse itself of its hopes. The litany of cooperative failure is long: the inability of the United States and China to mutually reduce their huge trade and financial imbalances (this is pre-Obama), which bears part of the blame for the global financial crisis and ongoing economic weakness (and is a nice example of the challenge of managing globalization in a multipolar world); increasing trade friction with China on IPR, subsidies, export controls, cybersecurity, currency pricing; aggressive Chinese behavior in South and East

China seas despite US efforts at mediation; Chinese and Russian obstruction of UN efforts at humanitarian relief in Syria; growing competition for resource access in the Arctic; Russian annexation of Crimea and its invasion of eastern Ukraine; and eroding Russian implementation of arms control agreements. Balanced against this is a more modest list of gains: chemical weapons removal from Syria; the maintenance of sanctions against Iran as the nuclear negotiations proceed; continued Russian support for the diminishing NATO presence in Afghanistan; and a United States–China bilateral climate deal. The balance sheet is still incomplete, but leaning strongly toward deficit, and the direction of change suggests sustained great power rivalry, with all its negative consequences for global problem-solving.

With specific reference to the pivot, Asia is as subject to sudden, discontinuous change as any other region, and the consequences of potential surprises are greater as our commitment to rebalancing deepens. Potentially shaky assumptions include the sustainability of growth in Chinese power (which touches on the viability of its growth model, quality of governance, popular demands for improved quality of life, regional stresses, and aging population and productivity issues); the extent of demand throughout Asia for an enlarged US presence and more explicit security guarantees; the viability of regional balancing strategies given the multitude of conflicting interests within Asia; our capacity to deliver on the strategy given budget constraints and congressional opposition to trade promotion authority; and downside events in those regions still critical to US interests (Europe, Middle East) from which resources have been drawn and which now demand renewed, and possibly expanded, US commitments.

Relative gains in Chinese power are easily demonstrated quantitatively but are based on extrapolations of economic growth rates and military spending that fail to reflect recently observed (and previously predicted) weaknesses in Chinese growth (and in its growth model) and quality/capacity of governance, not to mention longer-term demographic challenges. Asian demand for an increased US presence is heightened by recent Chinese assertiveness in territorial disputes, but tempered by our allies' need to preserve economic ties with China and their determination to avoid being forced into a choice between the two great powers. There are wide differences among Asian states in their attitudes toward the two great powers and toward each other that inhibit effective regional balancing against China and enable China to deal with each individual state from a position of strength. Asian doubts about the future capacity and willingness of the United States to guarantee their security are increasing as the US-China military balance, particularly in the region, shifts adversely. While encouraging a larger US presence in the region, their objective is not to draw the United States into a bipolar regional conflict over sovereignty and security, but to increase their own bargaining leverage with China over a range of economic and territorial negotiations. While fearful of China, they also favor US-China cooperation, as this expands their choices and minimizes the risks of collateral damage from great power conflict.

Expectations of stability and security in Europe, a seemingly safe assumption enabling a shift in priorities, now confront a weakened euro, anemic European economic growth, internal EU divisions both North-South (on internal migration, spending, institutions) and now East-West (on relations with Russia and the United States), and increased political extremism. Though many argue that a stable Middle East balance will develop in the aftermath of US retrenchment, the civil war in Syria and its spillover into Turkey, Lebanon, and now Iraq, the rise of ISIS, Iranian nuclear developments and regional reactions, and renewed Israeli-Palestinian violence—all will guarantee continued American engagement, preferably built into our global strategy, or in reaction to events.

The greatest source of uncertainty about the future of Asia is, of course, China itself. The operating assumption is its continued rise, as robust economic growth enables strengthened military capacity and fuels greater ambition, both in the region and globally, in military strategy, diplomacy, and economics. Yet China is confronted with multiple internal and external limits to the sustainability of its growth model and hence to its legitimacy. These challenges include a falling growth rate partially disguised by over-reporting of production and government investments in unneeded capacity; diminishing external demand for its exports, and reduced tolerance for its mercantilist trade, currency, FDI, and technology practices; persistent poverty and income inequality; its rudimentary safety net and an aging population; rampant corruption and official lawlessness; the Communist Party's tenuous authority in the provinces; abuses of the physical environment, and high susceptibility to disasters both natural and man-made. Popular anger at these conditions is increasing. Given our tendency to exaggerate the capacity of other great powers (the Soviet Union throughout the Cold War, Japan in the 1980s), the gap between our assumptions about China, and emerging realities, may be quite wide.

The pivotal issue here is political, namely whether the CCP has the will and political capital to rebalance its economy in a way that virtually all observers and many party leaders acknowledge is necessary. The groups that have grown rich and powerful around the investment- and export-driven growth model are not about to embrace reform, and even without political opposition, the challenges of rebalancing the economy, without deepening the growth crisis, are formidable.

Successful rebalancing would lead to a reduction in China's trade surplus, returning greater stability to the international system and reducing the likelihood of protectionist policies being pursued by the Americans and the Europeans. Should China fail to meet these challenges, plausible scenarios that invalidate much of the conventional wisdom about China are easily imagined. In our scenario work at the Center for Global Affairs at NYU, we constructed three alternate futures that turned on how the CCP responds to these challenges: "Partial Democracy," in which an underperforming economy generates growing demands for political reform from below, which the party accommodates; "Strong State," in which the party responds successfully to demands for economic rebalancing, while maintaining its legitimacy and absolute authority; and "Fragmentation," in which both economic and political reform fail, the party becomes immobilized by internal divisions and fear of growing unrest, and the provinces take on greater authority.

China is, of course, not the only source of rapid change and potential future surprises in Asian—note the recent unexpected economic and political changes in Myanmar, the reinterpretation of the Japanese constitution to permit a more robust defense capability and strategy, re-eruption of historical animosities between South Korea and Japan, the turn for the worse in Chinese conflict with the Philippines and Vietnam over the South China Sea Islands, and with Japan over East China Sea islands, and an ever-surprising North Korea. The region as a whole is diverse, collaborating more closely in some areas, but still in open conflict in others, and with divergent threat perceptions of China and of the role the United States should take in the region. It is highly vulnerable to natural disasters and to external economic shocks (though with large financial reserves, less so than historically). Future wild cards abound, from regime-threatening weather events, to bilateral conflicts, to sudden impulsive shifts in North Korean strategy.

## The Syrian Civil War

If Iraq is the graveyard of the Bush Doctrine, Syria may be the resting place of "restraint." The reluctance to intervene forcefully was clear from the outset, but in the early stages of the civil war was easy to justify, based on what turned out to be wishful thinking that Assad's demise and the triumph of the secular opposition were just matters of time. What the administration clearly failed to consider was Assad's large, untapped military capacity and his open arms and financial pipelines from Russia and Iran. With US promises of assistance to the opposition being slow-walked through the bureaucracy, the balance on the ground tilted toward Assad and his allies, and the administration's restraint came to constitute acquiescence in Assad's preservation. This result, acceptable—though unpalatable—within a pragmatic, stability-first strategy, has been greatly compounded by the eclipse of the secular Syrian opposition, the rise of ISIS, and the renewed fragmentation of Iraq, unmitigated strategic disasters that now incentivize at least implicit cooperation with our principal adversary in the region, Iran, in order to save Iran's closest allies, the Shiite government in Iraq and Assad himself, and provides Iran with potential leverage in the nuclear negotiations. There's no better example of how kicking the can down the road leaves only bad options.

Much of this was foreseeable, although not in its precise details. At CGA/NYU we did a Syria scenario exercise in February 2013. Reproduced below are the introductions and conclusions from the three alternate scenarios for Syria 2018; not included are the detailed narratives. As in the case of the Iraq exercise, the "Regionalized Conflict" scenario was right on target, imagining escalation, spillover, and radicalization very much as events played out. The "Negotiated Settlement" scenario was deemed best but also least likely, providing some cover for US restraint in Syria, and thus perversely becoming an enemy of the good enough, namely containment.

---

### SYRIA SCENARIO 1: REGIONALIZED CONFLICT

This scenario outlines how the sectarian, ethnic, and geopolitical cleavages that have roiled Middle East politics for decades are further widened by the escalating stalemate inside Syria. As with the "Contained Civil War" scenario below, outside actors furnish arms and funds to their respective clients at a level sufficient to safeguard their interests, while trying to avoid direct engagement in the hostilities. In this "Regionalized Conflict" scenario, however, events on the ground trump these calculations, rendering a strategy of calibrated violence unfeasible. Escalated fighting fans sectarian hatreds and murderous campaigns against civilian populations. As the country fragments along sectarian and ethnic lines, attitudes harden and hopes for a negotiated settlement are doused. As regional players respond in kind to shifting dynamics in the fighting with increased support to their faction of choice, momentary advantages enjoyed by one side are eventually countered by others, producing an uncontainable stalemate.

A regionalized conflict is, in fact, already well underway, with regional powers contending for influence through (imperfectly controlled) proxies, sectarian violence and refugee outflows threatening the fragile sectarian and ethnic fabric of nearby states, and increasing armed clashes across borders. While the "Contained Civil War" scenario imagines how these forces might be managed, this scenario suggests how inherently tenuous such a "managed" stalemate would be.

## Syria Scenario 2: Contained Civil War

This scenario outlines how an increasingly violent and protracted civil war remains relatively contained within Syria. Outside actors furnish arms and funds in an attempt to promote their interests within the region and to balance the moves of their adversaries. This support is provided to various factions within the opposition and to Assad's regime. It has the effect of fueling the violence within Syria, with frequent spillover across borders. Unlike the "Regionalized Conflict" scenario, however, it does not escalate to direct, sustained conflict between regional actors. The Syrian civil war instead settles into a protracted, multisided sectarian conflict with aspects of proxy war among regional rivals. This "containment" is a result of the self-interested calculation among great powers that unrestrained support to favored factions will produce diminishing returns, and among regional actors that their own internal political stability is imperiled by continued support for Syrian factions. These informal restraints limit the risk of regional conflagration, permitting the proxy war to continue largely unabated. They do not, however, lead the conflict toward resolution in the form of either military victory for one side or a negotiated settlement based on compromise. With a fractious opposition, increasing radicalization and heightened sectarian violence, persistent fighting across borders and unsustainable refugee flows, the long-term plausibility of "contained civil war," without some formal understanding among outside actors or a political settlement among internal rivals, is very much open to doubt.

## SYRIA SCENARIO 3: NEGOTIATED SETTLEMENT

This scenario begins, as others do, with the current stalemate, competition among outside actors for influence in Syria, all sides with options for escalation, growing territorial fragmentation, and political radicalization. Getting from here to a negotiated settlement presents obvious challenges. The easiest paths to serious negotiations would be decisive military victory for one side and an essentially imposed settlement, or a de facto territorial partition ratified and maintained by international agreement. Decisive military victory, however, is unlikely (despite recent regime gains), as each apparent advantage is eventually matched by escalation from the other side. The complexities of a sectarian settlement, the resilience of the regime, and divisions among the opposition and among outside powers militate against an enforceable partition. Thus, this scenario depends upon a subtle and potentially transitory shift in the power balance that creates sufficient incentive for most parties to negotiate, and that enables timely outside pressure toward this end to be effective.

Internal Sunni conflict between the Western-backed Free Syrian Army and Islamist militias prevents the opposition from translating the regime's weakness in manpower into sustainable military or political advantage. With neither side able to pursue a decisive military victory, the international community has a momentary opportunity to leverage negotiations between Syrian actors with international participation. This would most likely require Security Council consensus on a ceasefire, interim territorial arrangements, an international presence, and a process of talks between a unified (but probably unrepresentative) opposition and a post-Assad Baathist regime. Reduced-scale internal violence continues, but the tenuous ceasefire holds, and the outlines of a political and territorial settlement begin to emerge.

Spoilers abound in this scenario. A major US and European role in brokering such a deal rules out a seat at the table for Jabhat al-Nusra, who would be incentivized to continue the fight and would have to be blunted on the ground through ongoing counterterrorism operations by outside powers, a task they are currently unwilling to assume. The role of other radical Sunni militias is problematic, and these groups' external supporters would have to see benefit in restraining them in the interests of regional stability and their own security. Hezbollah is a major challenge, but in this scenario is looking to cut its losses and refocus on core interests in a fragmenting Lebanon. Iran would have to be convinced that both the Syrian regime taking shape and the envisioned territorial settlement advances its security needs better than open-ended support for Assad. Russia, too, would have to acknowledge these same realities, with assurances that its interests are respected in an opposition-governed Syria.

## Conclusion: American Policy Choices across Alternate Scenarios

The current American approach of incremental and reactive escalation—exhibited most recently in a commitment to provide small arms and ammunition to vetted opposition forces, as well as diplomatic efforts to organize a peace process—cannot be assessed on the basis of Syrian / Middle East scenarios alone. Asian and domestic priorities may demand restraint in Middle East policy regardless of how many people die or are displaced inside Syria, how rapidly and extensively the conflict spreads, or how many red lines are crossed in the process. The value of the scenarios taken together is not that they dictate policy responses, but that they suggest the shifting balance of costs and benefits associated with alternate courses of action, as the crisis continues to transform the region. Their purpose is to take both wishful thinking and despair out of the policy debate.

The "Regionalized Conflict" scenario depicts a downside future that was taking shape as the workshop assembled in February 2013 and has gained in plausibility and gravity since then. It should be thought of as the region's most likely future, given the enlarged role of Hezbollah and the fears this generates in Israel, the deepening refugee crises in Lebanon and Jordan, escalating sectarian conflict in Iraq, and expanding commitments of outside powers—Iran, Russia, the EU, the United States—to materially support their local allies. The scenario depicts massive humanitarian effects (as of this writing, deaths are estimated at just over 100,000), reignited civil wars in Lebanon and Iraq, a destabilized Jordan, Turkey preoccupied with internal security and increasingly using force to protect its borders, and escalating direct clashes between Israel on the one hand, and Hezbollah, Syria and Iran on the other. Increasing tensions between the

United States and Russia, already evident despite the measured US response, is also a feature of the scenario, and the deepening/broadening of the conflict further exacerbates these tensions. The scenario also questions how long China can remain aloof from a conflict in a region of growing importance to its economic and strategic priorities. This suggests that the Obama administration's Asian strategy is at risk unless the Syrian conflict can be contained or resolved.

Can US interests tolerate this degree of regional turmoil? At what point will the potential damage to regional stability and great power relations outweigh the risks of more direct participation inside Syria? Is the United States capable of protecting its Middle East interests—defense of its allies, prevention of terrorist safe-havens, WMD nonproliferation, containment of Iran, security of oil flows—as the regional map is violently redrawn? Can it do so if the Russia/Iran/Hezbollah/Assad coalition prevails? If other, less damaging outcomes are plausible, as our "Negotiated Settlement" and "Contained Civil War" scenarios suggest, how can the United States deflect the present course of events, which are clearly on a worst-case trajectory, toward these more favorable scenarios? Can it up the ante sufficiently to generate greater leverage with both allies and adversaries without contributing to the destabilizing outcomes it most fears? It is clear that the metastasizing Syrian civil war is a game changer for the region, and possibly for the global system. The management of these shocks is hard to imagine without the presence of American power.

Whether or not the United States acts robustly to prevent the worst case, the scenario requires that we imagine US interests and policies in a violently collapsing Middle East. State fragmentation and a regionalization of sectarian conflict may entail some upsides for the United States, and the direct effects of regional turmoil on American security and prosperity might be minimized. An extended regional conflict between Sunni and Shiite extremists may sap the energy and legitimacy of both; Iran may find the job of managing its interests in a collapsing Syria to be a drain on resources and dangerously provocative to Israel; expanded US support to local actors whose interests are more or less in alignment with America's could produce an effective regional balance of power without America's direct participation in the conflict. The surge in American energy production and exports may permit a somewhat more detached posture toward Middle East instabilities. The collateral damage to great power relations depicted in the scenario might be contained by formal understandings to avoid worst-case outcomes. While the scenario offers no guarantee that events, large-scale use of chemical weapons being the most obvious, won't demand intervention, the conditions described above may be deemed "good enough" to allow Asian rebalancing and domestic reform to continue unimpeded.

But this is mostly wishful thinking. Without US leverage to shape events, adversaries will be emboldened. Allies will protect their interests by cutting deals, expanding support for Syrian factions of choice, or, if they have the capacity, intervening directly, possibly resulting in state-to-state conflict in the region and inviting intervention by great powers. The humanitarian crisis will deepen. The already shaky global economy will suffer as the security of energy trade is compromised. Permanent damage will be done to both the region and the structure of international stability and accountability.

If the worst case is intolerable and a negotiated settlement (the third scenario) is highly unlikely, the "Contained Civil War" scenario considers whether regional dynamics and the policies of external powers can work together to contain the damage.

As bad as Syria has become, the first scenario demonstrates how much worse it could get. As regional and global actors face the impending reality of a collapsed Middle East, might they, with some encouragement, become more amenable to strategies that require constraints on their behavior? Would these constraints be self-imposed, or would they require some formalization? Could the United States contribute to this outcome, not by hoping that others imitate its own "light footprint," but by imposing higher costs on those feeding the conflict, to the point that their calculations change and opportunities are created for damage limitation?

A contained civil war does not require a settlement (see scenario 3), but a settlement might result from a combination of stalemate on the ground, higher levels of violence that deplete the combatants, and restraint among outside powers based on their realization that worst-case risks are increasing, while the returns on their support of favored groups are diminishing. A commitment to this outcome would reflect outside powers' sense of hopelessness that the fragmented opposition and an intransigent regime can ever be brought to the negotiating table and that containment is the least bad plausible outcome, with some promise of avoiding a regional sectarian conflict. Thus, the violence continues, ameliorated by increased international humanitarian assistance, diminishing over time by informal understandings to limit external arms supplies. The damage already done to Lebanon, Jordan and Iraq, and to relations among the region's states, will continue to burden the Middle East for years to come, but the violent dissolution of the post-Ottoman system is prevented, and the negative fallout for great power relations is limited. The containment is not complete—much too late for that—but the hemorrhaging beyond Syria and its immediate neighbors is arrested.

This option, and the policies necessary to achieve it, is often lost in a two-scenario debate between the advocates of forceful intervention and a negotiated settlement brokered by the United States and cooperating great powers. The former approach runs the risk of further escalation and entrapment, while the latter encourages US restraint as others are creating new realities on the ground.

Were the United States to adopt containment as the optimum choice, what specifically would this entail? As the scenario narrative argues, Russia would have to tire of its ongoing commitment to an intransigent Assad in an endlessly stalemated civil war. It would have to see opportunities for securing its naval interests in a post-Assad Syria, fear spillover of the conflict into its own restive regions, and experience real costs associated with growing tensions with the United States. It would then have to be prepared to join the United States in restricting its own and others' support to the combatants. There is no indication at present that Russia is close to drawing these conclusions. With Assad consolidating his hold on western Syria and the United States committed to gradual and contingent escalation, there are no forces at work to change its views. Again, US leverage is essential if we are to avoid worst-case scenarios, whether this leverage is applied to brokering a settlement (scenario 3) or to engineering a stalemate that Russia and others understand will not succumb to their next escalation.

The temptation, if the logic of this scenario is accepted, might be to view an imperfectly contained stalemate, achieved by Great Power agreement to starve the conflict, as a prelude to a negotiated settlement. However, the scenario narrative's end game suggests that trying to push the internal combatants toward political compromise would further destabilize Syria and possibly the region. The regime is unyielding, and has the internal and external sources of support to remain so. The opposition is fragmented, and any serious push toward negotiations will arouse the spoilers. All view the conflict as existential. All sides have regional allies who would fear marginalization in a negotiated settlement. Better to invest in a more modest, but attainable and useful, outcome. This would reflect the sad conclusion that Syria's nightmare will continue, even that Syria's days as a sovereign country are numbered, but that a contained civil war is the best we can do, given the lack of internal and regional consensus.

The "Negotiated Settlement" scenario represents the most plausible version of the US administration's strong preference. The difficulties are evident for all to see: a still powerful regime fighting for its survival, with both internal and external support; a fragmented political and military opposition, with some factions amenable to political solutions, but others violently opposed; regional actors deeply implicated in a proxy war in Syria; and Western states eager to broker a settlement, but unprepared to invest in additional leverage necessary to achieve it. These conditions, if allowed to continue, are among the least conducive imaginable to a negotiated solution. Indeed, they are more favorable to the first scenario or, if Great Powers choose to prevent the worst case, the second. What are the United States and its partners prepared to do to create preconditions for a serious negotiation? How can we work effectively and directly on the local antagonists, and on their external enablers, to move from fighting to, mostly, talking?

As presented in the scenario, the negotiations are facilitated by military stalemate along fairly well-defined territorial boundaries. This stalemate is maintained not by the aims of local actors, all of whom continue to seek dominance within Syria, but by the competitive balancing of regional and global players, whose support to all sides prevents the hegemony of any. In this sense, the stalemate is uncoordinated, dynamic, and fragile, subject to sudden reversal as internal or outside actors seek a decisive edge on the ground. It produces a disposition among outside actors toward negotiations only as they reach two seemingly contradictory conclusions: neither side will be permitted by other suppliers to gain a decisive edge, and further escalation threatens additional spillover and their own political stability/legitimacy. With no end in sight and the political, economic, and human costs mounting, it might be possible for great powers to broker a ceasefire and the beginning of a negotiating process leading toward some sort of territorial partition. This is not the political settlement that the United States seeks. Indeed, expectations of genuine power-sharing may be the enemy of this "good enough" arrangement that stems the bleeding and, if maintained by multilateral forces, could evolve toward genuine political compromise and a Syria with restored sovereignty.

Again, these conditions are not now present. External actors clearly have not concluded that there is a stalemate in Syria or, if one were re-established, that it couldn't be overturned to their advantage. The United States, reacting incrementally to a dynamic and uncertain battlefield and trying to calibrate its responses to reinforce the "good" opposition without risking more spillover and another Middle East quagmire, encourages its adversaries to hope that their next escalation will lock in advantages on the ground. While aiming for a stable stalemate and a war-ending deal, the United States will have to escalate more rapidly and decisively, even preemptively, in order to fundamentally alter the calculations of the regime's supporters. It will have to put other actors' worst-case scenarios in play if it is to gain traction in Syria.

Looking back from the summer of 2014 to these February 2013 scenarios, our pessimism about a negotiated settlement was fully warranted, and scenarios 1 (fragmentation / radicalization / regional spillover), and 2 (stalemated/contained civil war) are very much in play. We also pointed out that containment through informal pressure from outside powers would be tenuous, easily breached by events on the ground or competition for influence among those outside powers. That skepticism also appears warranted. Syrian fragmentation and regional spillover, the worst-case scenario in 2013, has emerged as the dangerous present reality, with more to come. Our conviction that enhanced American leverage was essential, not to win the war in Syria, but to create the prerequisites for containment or resolution, also seems borne out by subsequent events.

Chapter 3

# Value of Scenarios

Multiple scenarios are designed to challenge the mindsets policymakers bring into policy debates, by presenting alternate narratives that capture less conventional but plausible views of the future. They are not predictive, nor are they "actionable" in the sense that decision-makers can operate with high confidence based on their speculations. Quite the contrary: they are intended to make explicit and to deconstruct predictions that force-fit analysis to preferences or other forms of bias. In doing so, they can reveal dubious assumptions, conveniently overlooked policy trade-offs, and plausible future "wild card" events and trend developments that can invalidate current policies and pose new challenges. They can open up minds to alternate ways of interpreting available intelligence, retarget intelligence to clarify and narrow new uncertainties, and make decision-makers more receptive to early warning signs of new trends. They can help decision-makers overwhelmed by noise get out in front of the rush of ongoing events, by discerning and focusing on the signals and testing policies of prevention or encouragement. By engaging the future, alternate scenarios avoid the equally hazardous extremes of infinite uncertainty, which can produce paralysis or willful disregard for what we do know,[1] and excessive self-confidence, with attendant risks of ill-conceived actions and unintended consequences.

Multiple scenarios are constructed because our goals are insight and improved recognition, not replacement of the prevailing paradigm with a new conventional wisdom equally subject to degradation. Consider the event-driven and transitory character of post–Cold War intellectual commentary: the decade of the 1990s, variously described as threatless, flat, global, the end of history;[2] the following half decade, as pervasively insecure, a perfect storm of terror, WMD proliferation, rogue and failing states; and the more recent emphasis on declining relative American power and emergent multipolarity. This recent contender for intellectual primacy spans observers as disparate as Robert Kagan, Fareed Zacharia, John Ikenberry ("In the decades to come, America's unipolar power will give way to a more bipolar, multipolar, or decentralized distribution of power"), and the National Intelligence Council (see notes 3–6 of chapter 1). This consensus is reinforced by the global economic crises, which in the dominant view accelerates the erosion of unipolarity[3] but could cut in the opposite direction, undermining governance and power in "rising" states (and regional institutions) now experiencing reductions in growth, and reinforcing those states with strong and diversified economies, robust civil society, and legitimate political institutions capable of responding to economic stress. Indeed, expectations of a multipolar future have provoked a reaction among liberals and neocons, who predict American renewal and restored primacy.

The principal objective of alternate futures is to improve observation of a rapidly changing and complex reality, and to encourage early recognition of and reaction to emerging trends that may shift the ground under current policies. The scenario conversation is informed by theory, but not committed to a single paradigm. Instead, it is consciously designed as a dialogue across theoretical boundaries, disciplines, and cultures. Participants in scenario-building workshops are encouraged to step back from their assumptions, exercise their imaginations to envision new trends from fragmentary data, invent "wild card" events that combine plausibility with impact, and construct complex narratives that cumulatively make a convincing case for a future that may at first blush appear inconceivable, or too undesirable to contemplate.

The process avoids the extremes of groundless speculation on the one hand, and categorical, theory-driven pronouncements on the other. It emphasizes causation and explains scenario conditions as consequences of multiple "drivers" interacting in sometimes surprising but believable ways. By leveraging knowledge (not everything is uncertain) without exaggerating our foresight, the process can narrow uncertainty, reduce surprise, eliminate implausible futures, identify policies that "work" across a range of future conditions, and help policymakers manage risk.

We can group the value of alternate scenarios into three broad categories. *Deconstruction* is essential if the alternate scenarios emerging at the end of the process are to take us beyond minor variations on the official future. Given that many surprises reflect willful blindness, a well-designed and facilitated scenario exercise adds value by forcing assumptions into the open and subjecting them to scrutiny. The disaggregation of the scenario subject (country; issue) into component drivers of change and their interaction, produces a bottom-up unpacking of often simplistic explanations of shaping forces.[4] Proposals for alternate scenarios from participants or the facilitator provoke debate about the sustainability of prevailing assumptions, and the plausibility of alternate, less extrapolative or preferable possibilities. The discussion of drivers and possible futures often reveals a fundamental lack of consensus about the present, emphasizing just how constructed and motivated are our versions of current reality. Well before the actual construction of alternate scenarios, the discussion of drivers, potential scenarios, and wild cards begins the process of assumption scrubbing and opening of minds to alternate possibilities and facilitates learning.

I'll call the second category *improved recognition*. Scenarios can provide alternate contexts for observing and explaining the flow of events. They alert us to the meaning of events that can appear random, anomalous, and inconsequential, if they are observed at all. Any global strategy sets up priorities for intelligence collection, as well as expectations for what we will discover as we observe and analyze the facts. Thus, what we deem relevant, and how we evaluate this already self-limited set of data, is subject to error, and the magnitude of potential error widens as strategy sticks to the straight and narrow and collection priorities follow. In the fast-changing world described in chapter 1, the widening gap between what we think we know, and what we need to know, can become disabling, guaranteeing strategic surprise and depriving decision-makers of the opportunity for course corrections necessary to preserve the essence of a chosen strategy.

Alternate scenarios can improve recognition in several ways. They can highlight assumptions underlying current policy that have lost their robustness, thus suggesting a closer look at sources of change. They can jog our images of the future, which can easily bias in favor of current policy, or national/cultural identities, and by doing so can open our eyes to alternate evaluations of the significance of unfolding events. They can provide a context for early warnings of specific threats, lending these warnings greater credibility. They can identify indicators of seemingly unlikely but plausible and consequential futures, thus suggesting new intelligence targets and enabling systematic tracking of these potential developments. And, most ambitiously, they can suggest alternate assumptions about the world, as it's becoming, that are more robust than the assumptions underlying current policy.

The third benefit is *preparation*. Deconstruction and improved recognition are essential to achieve better preparation: the process must break down "grooved thinking," then suggest alternate ways of imagining and tracking the future, before we reach the ultimate goal, which is better preparation for a very challenging world. But this goal must be kept in mind throughout the process, if the benefits are to extend beyond improved insight. Peter Schwartz and Doug Randall have written that "anticipating strategic surprise is ultimately valuable in terms of preparedness. Organizations that have thought about such significant issues are much more likely to discern important, emerging trends early on; identify the indicators that tell them something big is happening; and put in place the sensors to detect strategic surprise as it unfolds. If the key indicators are getting worse, the worst-case scenario becomes more and more plausible. This in turn gives organizations the ability to act in advance if they believe a particular scenario is unfolding. It gives them more maneuvering room and time to create new options."[5]

Improved preparation is more a matter of thoughtful anticipation of, and reaction to, change than of formal planning and forecasting. Andrew Erdmann speaks of the need "to focus new formal and informal mechanisms to bring insights from strategic planning to bear on major decisions wherever they are made—in the boardroom, in the hallway, or on the eighteenth green."[6] Although we should aim at anticipation and thus minimization or prevention of surprise, we should also prepare ourselves to take full advantage of those surprises that will, inevitably, occur. This requires that we react quickly and accurately to shocks, and doing so is far more likely if we have already simulated and evaluated alternate responses to hypothetical but plausible "black swans." Otherwise, impulsive, ill-considered actions, or immobility, will often result.

Policy testing (Are my policies working now? How might they degrade? How will I know? What incremental or fundamental policy changes might I consider to protect my interests in alternate future worlds?) is essential to risk management. The more global our interests, and the more change in our external environment, the more quickly current policies will reach diminishing returns, and the greater the risks that must be confronted. Some of these risks are associated with choice of grand strategy, as other actors respond to what we do, or as the overall environment changes (how could my overall all approach to the world go wrong? How can I anticipate, control, or manage these risks to strategy? What alternate choices do I have?); other risks will arise from shifts in the drivers of change, from technological to political. Preparing for these risks, by imagining how strategy might atrophy, then thinking through our responses, is the central value proposition of alternate scenarios.

Maximizing preparedness requires that alternate scenarios become part of a process of continuous future reality testing against current strategy and the assumptions underlying it. The scenario construction process itself (see next chapter for details) ideally jump-starts or reinforces dialogue among experts and policy or strategy makers that demonstrates value for the policy community: in the more rigorous and open treatment of extant assumptions; in the analysis of driving forces; in unfamiliar but plausible pictures of changed conditions; in the testing of current and alternate policies. In well-designed and well-conducted scenarios sessions, the quality of the conversation and the results demonstrate the tangible value of alternate futures and generate political capital to invest in a continuous process. Erdmann observes that "carefully prepared dialogue about emerging trends and their implications among executives is a much more powerful way to 'prepare minds' than yet another series of PowerPoint slides prepared by and for strategy staffs."[7] Greg Treverton, a senior analyst at RAND and new director of the National Intelligence Council, suggests that "policy makers and strategic analysts should seek to better coordinate their efforts . . . the products resulting from long-term analysis should be in the bloodstream of policy makers. They should be absorbed horizontally and vertically in the decisionmaking process."[8]

The premise is that hypothetical but plausible scenarios, and the process of constructing and revising them, can help us make decisions that prepare for, and in some cases shape, the future, without committing too deeply to a single view. We must plan strategy, invest in capabilities, prioritize intelligence collection, and communicate with stakeholders. All require operating assumptions about the future, which should be based on the best of current knowledge, intelligence, and analysis. The future I constructed in the first chapter is, in effect, a "most likely" scenario that lays out a broad context for thinking about strategy and policy. A strategy process informed by alternate scenarios accepts the most plausible assumptions we can make but establishes other, less likely but plausible conditions, then maintains a close watch on extant assumptions as current strategy is implemented and the future unfolds.

## What about Prediction?

But should we adopt a more ambitious objective, of predicting trends and events? My view, a product of observing increasing complexity and frequent surprise, is that the potential damage from overconfidence in our own predictions is greater than the benefit of acting based on marginal improvement in forecasting accuracy, whether achieved through "big data" or through polling of experts. If the world is becoming as chaotic as I've argued, the objective of observers hoping to inform strategy should be recognition of emerging trends and events, in preparation for change; policy agility, flexibility, adaptability are the attributes most in demand. Good strategy process trumps good predictions and strategic plans. Predictions tend to be extrapolative. They can achieve a degree of accuracy but sacrifice utility because they miss the tipping points, the big, strategy-invalidating events. They run the risk of generating actions based on excessive confidence and of blinding us to developments offstage. They can produce an illusory sense of comfort, when more often than not we need discomfort to maintain the required degree of alertness. Nate Silver writes that "one of the pervasive risks that we face in the information age . . . is that even if the amount of knowledge in the world is increasing, the gap between what we know and what we think we know may be widening. This syndrome is often associated with very precise-seeming predictions that are not at all accurate."[9]

Philip Tetlock, a political psychologist at the University of Pennsylvania, has written extensively on expert prediction, in his book *Expert Political Judgment*,[10] and as part of his ongoing Good Judgment Project, funded by IARPA (Intelligence Advanced Research Projects Activity). In the book, Tetlock reports on the predictive performance of experts he groups into two categories, based on contrasting cognitive styles. The categories are borrowed from Isaiah Berlin: hedgehogs know one big thing, foxes many small things. Hedgehogs are self-confident about what they know, committed to a parsimonious, theory-driven view of the world, and categorical in their predictions; they explain causation, invoke the lessons of history, and, in the face of observed reality that contradicts their predictions, urge patience. Foxes are more cautious in their views of the future, see multicausality behind events, try to apply knowledge from a variety of disciplines, are more open to debate and learning, and are less given to top-down explanations and conversation-ending pronouncements. They are self-doubting. Their thinking style, in Tetlock's view, is better adapted to a "rapidly changing world."

Foxes outperform hedgehogs in prediction contests. Tetlock reports that "good judges tend to be moderate foxes: eclectic thinkers who are tolerant of counterarguments, and prone to hedge their probabilistic bets and not stray too far from just-guessing and base-rate probabilities of events."[11] "The foxes' self-critical, point-counterpoint style of thinking prevented them from building up the sorts of excessive enthusiasm for their predictions that hedgehogs, especially well-informed ones, displayed for theirs. Foxes "'overpredicted' fewer departures, good or bad, from the status quo. But foxes did not mindlessly predict the past. They recognized the precariousness of many equilibria and hedged their bets by rarely ruling out anything as 'impossible.'"[12]

This cognitive style is fairly close to what should be sought on a collective basis in the interaction among experts and policymakers during scenario-building sessions. The problem with individual foxes as prognosticators, however, is that, while they do better than hedgehogs, "foxes are not awe-inspiring forecasters: most of them should be happy to tie simple extrapolation models, and none can hold a candle to formal statistical models."[13] Tetlock reports improved forecasting results in his current work, the Good Judgment Project, in which a group of "superforecasters" has consistently, over time, outperformed the mean in predicting specific events. We'll have to wait for the formal report on this work before drawing conclusions.

There is another issue with these forecasting results, however, which places "forecasting accuracy" in context of the complex, rapidly changing world that I've depicted in chapter 1: what kind of analytic skills do decision-makers need access to in a world perpetually remaking itself? Extremely well-informed observers who think like foxes (the characteristics of superforecasters) and are single-minded in their devotion to prediction consistently outperform extrapolative models, not to mention hedgehogs, in accurately predicting specific events. But these are averages, calculated over a multitude of predictions. What about the singular sea-changes, the revolutions, the breakthroughs, the "black swans" that come from outside our experience, for which a granular understanding of reality is of limited avail? What about the Arab Springs, the Russian invasions of Ukraine, the sudden emergence of ISIS? Shouldn't we consider which—if either—cognitive style is most likely to anticipate the big, current-strategy-undermining events, or at least make a sufficiently plausible case to get us thinking and preparing? Are not some forecasts accurate, but useless? Are some forecasts wrong, but useful? If we value-weighted the forecasting scores of foxes versus hedgehogs, who would emerge on top?

Tetlock acknowledges that one source of fox superiority is that they forecast conservatively: "Foxes do better because they are moderates who factor conflicting considerations . . . into their final judgments,"[14] issuing fewer extreme forecasts. With respect to potential political instability in Saudi Arabia, for example, foxes tend to favor "muddling through" scenarios, which may produce higher forecasting scores, but also, if fully incorporated into strategy (bet on the regime), builds up risk and increases the likelihood of ill-preparedness for unpleasant surprises. Tetlock also sees value in hedgehog-like self-confidence. "Hedgehogs made many mistakes, but when they were right, they were very right. When stunning discontinuities took almost everyone by surprise, it was a good bet that a few hedgehogs would be left standing to take credit for anticipating what no one else did. . . . Experts who in 1988 predicted the collapse of the USSR in 1993 might be forgiven for 'overpredicting' the collapse of other regimes."[15]

In a remarkable, and inadvertent, example of how fraught is the concept of predictive "accuracy," Tetlock in his 2005 book reports interviews with two Ukraine experts, conducted in 1992, as an example of the hedgehogs "not knowing when to apply the mental brakes." The fox predicted that the Ukrainian leadership would not provoke Russia, and that Russia was not spoiling for a fight. The hedgehog "foresaw growing tension with Russia that would culminate in interethnic violence, a Russian energy embargo and military intervention to protect Russians, and the forced seeding of territory to mother Russia."[16] Tetlock concludes that "wariness of facile generalizations helped this fox forge an integrative set of economic and political expectations that were more accurate than that of most other specialists."[17]

One can easily understand Tetlock's enthusiasm for the fox's "more accurate" Ukraine forecast: most in accord with Ukraine's economics and politics in 1992, with a Western orientation offering real promise for growth and stability, and a weakened Russia in no position to prevent this. It was closest to an extrapolation. It still looked to be correct in 2005, when the book was published. And it issued from an inquiring intellectual style that Tetlock favors. Nor should one expect decision-makers to have premised strategy for Ukraine and Russia on the hedgehog's predicted discontinuity, with relatively little "current intelligence" to back it up, especially when everything about international politics—in 1992—seemed to be going our way. But the value proposition for alternative futures has very little to do with predictive accuracy; it is to position seemingly improbable but consequential outcomes on our radar, to prudentially consider how our favored future might go awry, to make us more attentive to often subtle indicators that perfectly reasonable expectations are becoming less realistic. In the Ukraine case, taking the hedgehog more seriously might have left us less flat-footed when the "discontinuity" became reality. It also appears that our hedgehog had the intellectual equipment to understand something about Ukraine and its external position that our fox did not—its fragmented identities and loyalties, its ineffective government, and its location on the border of an aggrieved and powerful state, with NATO membership for Ukraine highly unlikely—that enabled him to extend his view beyond the unipolar and transitory conditions of the 1990s.

The fox-versus-hedgehog contrast of styles is a well-known problem in attempting to anticipate fundamental change, often expressed as the unavoidable choice between missed signals (foxes) and false signals (hedgehogs). The intellectually conservative foxes expect business as usual, tend to underpredict change, and, because tomorrow is more often than not much like today, are usually right, but often unexceptional, and risk being devastatingly wrong. The intellectually bold hedgehogs, in deploying elegant theories of causation, leap ahead of the stream of events to predict big changes, confident that the future will eventually catch up with their forecasts. If they're wrong, which they often are, they can still be usefully wrong. The Ukraine prediction, made in 1992, was wrong until 2013, when it was revealed as brilliantly correct. The downside with hedgehogs, of course, is that they overpredict and are fuzzy on timing, are too self-confident to learn from their mistakes or to modify their views in light of new information, yet are often more effective in debate than the self-debating foxes.

Without declaring myself as inclined toward either style, in constructing groups of experts for scenario building, it's useful to have both. The opportunity to speculate about plausible long-term futures should free the foxes from risk-averse analysis and data-driven extrapolations. The explicit discussion of assumptions, and the invitation to imagine futures premised on alternate theories, should bump the hedgehogs off their well-worn paths. The interaction of experts and policymakers with contrasting cognitive styles, skill sets, and national identities should, if done right, produce both seriousness of purpose and intellectual playfulness, a collective willingness to entertain alternate points of view, and a commitment to build the best case possible for divergent futures. The overall style of the conversation should be foxlike (open-minded, inquiring, imaginative, Bayesian), but prepared like hedgehogs to entertain big ideas about the extent and direction of change.

## Timing Is Almost Everything

How to maximize the value of alternate futures for policymakers is, in part, a matter of timing. New administrations, eager to put their own imprint on grand strategy, sometimes to deliver on campaign promises, are primed for the basic rethinking about the world that alternate scenarios deliver. Richard Haass makes this point: "Policy planners have a structural advantage at such times because all new administrations feel a certain pressure to invent. Further along in any administration new ideas gain less traction because, with the passage of time, policymakers inherit existing policies. Absent intervening events that create windows for rethinking, changing policies in an administration tends to be like steering a supertanker: difficult, slow, and sure to meet resistance."[18] But Haass also observes that "some moments have been especially propitious because history demanded new ideas. Crises and major events came along that called for innovative responses, and there was a search for new paradigms. The September 11 attacks are a good example. After the attacks, policymakers were suddenly in the market for certain types of policies and ideas that would have received little or no support before. . . . Ends of great wars are another example. It is no coincidence that policy planning was conceived when it was, at the end of World War II. One geopolitical era had come to an end, and a new geopolitical era was dawning. . . . By contrast, in the 1960s and 1970s, most of the thinking in American foreign policy was contained within a paradigm—containment—in the context of the cold war."[19]

In all three of these favorable circumstances (end of a great war, a consequential and unpleasant surprise, a new administration) a surge in political leverage is combined with incentives for rethinking assumptions. Both are essential: the "after victory"[20] moment that concentrates power in the war-winning state and produces a determination not to repeat recent history; the increase in executive authority and sense of national identity that accompanies new threats and delivers political capital to leadership, as it provokes new ideas about how to respond; and the "honeymoon period" enjoyed, fleetingly, by new administrations, combined with a desire to do things differently.

Jeffrey Legro has written about conditions that facilitate fundamental changes in grand strategy, and how the process of change works. In *Rethinking the World*[21] he proposes a conceptual model in which a state experiences an external shock that appears to nullify existing strategy. Note that incremental erosion in strategy effectiveness is deemed insufficient to precipitate such rethinks, given the political and cognitive demands of fundamental strategy change. The experience of shock generates debate within the policy community, and more broadly among observers and the public. The normal, relatively narrow set of credible ideas, now with less legitimacy, broadens to encompass previously marginalized ideas arguably better suited to changed conditions. The debate that ensues has indeterminate outcomes: not all shocks produce a revolution of ideas, nor do those ideas that do rise to the top necessarily work. As in any policy debate, even in democracies with competitive elections, robust civil society, and lots of intellectual capital on all sides, the playing field is not level, and resistance to change remains. But the possibility that new ideas will emerge is much greater in these circumstances, and policy entrepreneurs have a real chance to sell something new. What follows, in Legro's model, is a period of experimentation, when new concepts are applied and the returns on this investment are assessed. Success confers legitimacy and consolidates a new paradigm (containment); failure (the Bush Doctrine) loses elections and sends us back to the drawing board or to conventional doctrine.

## Beyond Timing: The Importance of Process

Given the contingent and uncertain shape of emergent reality, and the goal of creating value for policymakers, the durability of any particular set of alternate scenarios will be limited. Large, formal scenario studies have their place, especially those focused on the more distant future—the four NIC reports are examples. But by and large, the scenarios will begin to lose their direct relevance to policy unless embedded in an ongoing process of hypothesis testing and policy reevaluation against incoming information. New data and insight may broadly confirm one scenario, invalidate another, suggest new scenarios to consider, reveal impacts of previous policy decisions—either intended or not—call for additional research or collection to clarify uncertainties, and alter the mix of policy options. In the best of circumstances, this way of thinking about policy occurs spontaneously: the formal scenario process provides greater discipline and structure to policy debates, thus increasing the likelihood that such thinking will be employed. The capacity should ideally be sited in the policy agencies themselves, and at the NIC, with extensive "outside" participation. The National Security Council, the National Economic Council, Policy Planning at the Department of State—all could assemble the right mix of expertise, imagination, and policy savvy needed to do this well, though as explained in the final chapter, the NSC should lead the effort.

For those interested in improving policy process and outcomes through scenario approaches, the lesson here is to be opportunistic: surprise (or victory in war, or a new administration) may afford an opportunity for long-overdue rethinks of assumptions experiencing slow and unnoticed collapse. But to take advantage, analysts must track policy demand for the approach, cultivate policy supporters, think ahead of policy clients about potential shocks, be ready with policy analysis relevant to a changed environment, and be prepared to offer alternate strategies better suited to the emerging world. They must be sufficiently independent of policy clients to entertain discomforting assumptions and politically incorrect responses, but close enough to understand how to market new insights.

Thus, to be effectively opportunistic, the process of "rethinking the world" must, at some level, be continuous. To be ready with the right product at the right time, those doing alternate futures must not be as reactive as policymakers, but must imagine potential surprises and be ready to meet new demands for explanation and prescription. If such a process has high-level policy support, there is a decent chance that unpleasant surprises can be leveraged to produce a new and sustainable strategy paradigm. And if I'm correct about accelerating change, diffusing power, and multiple actors of consequence, no episodic scenario process or set of planning documents can possibly provide the heads-up so essential to defense of US interests.

But what about minimizing surprise? Haass, Clark, many others are skeptical that scenario-like thinking can find leverage outside the exceptional circumstances of new leadership, end of great wars, or major shocks. That leaves us with a problem, namely an erosion of policy effectiveness, slowly diminishing returns that don't shock us out of complacency, but if measured cumulatively suggest a changed world, impending surprise, and a need for new approaches. Indeed, the surprises that do generate demand for basic reassessments are usually preceded by long periods of incremental change unnoticed and unmeasured by most observers. Let's remember, not everyone was surprised by the collapse of the Soviet Union, or 9/11, or the Russian annexation of Crimea. The fact of—occasional—accurate anticipation of apparently discontinuous events suggests that an ongoing process of alternate scenario construction and reconstruction can allow us to track forces for change and anticipate what would otherwise become major shocks.

Also important, but again not in demand until often too late, is risk assessment associated with new policy initiatives or grand strategy innovations. We should always be asking what could go wrong, not to delay closure on new policy, but to sustain it by managing the risk associated with new priorities. This prudential value of scenarios is relatively greater today than during periods of bipolarity or American primacy. We are more vulnerable to outside events than during the Cold War (deterrence and containment worked, and economic interdependence produced less vulnerability than globalization). "Planning" is less relevant because we don't shape our environment as we were able to do during the high point of "policy planning" in the late 1940s and 1950s. Rationally allocating scarce resources, getting early warning of impending threats, evaluating the costs and benefits of existing and proposed policies, reacting appropriately to surprises once they happen, are now more important as sources of demand on analysis, intelligence, and forecasting. Agility, adaptability, resilience are essential determinants of success in strategy. Anticipation itself becomes a source of power, a force multiplier, a competitive advantage, when other, material sources of power are in relative decline.

## What's a Good Set of Scenarios?

If utility is the ultimate goal, *plausibility* is the first essential requirement. Plausibility in scenario construction does not mean probability or likelihood. It entails a subjective judgment from those who construct the scenario, then by the reader/user, that each alternate scenario makes a convincing case that the narrative could unfold more or less as described, that the endpoint of the narrative, regardless of how different from present conditions or inconceivable at first blush, is believable. This requires that the alternate narratives flow naturally from present circumstances, then deflect onto separate paths as plausible future events occur in one scenario, but not in others. In walking the reader into the future, the scenarios should explain themselves, describing the flow of events, but periodically step back from the narrative to highlight drivers and events that are shaping the story. That is, they should explain causation as best we can, based on current knowledge about the pace of technological innovation, the impact of relative declines in income on political stability, the political economy of the specific country in question, and so on. This transparency about driving forces and discreet events enhances the credibility of the narratives, facilitates learning, keeps the debate about the future going, and enables policymakers to assess and exert leverage over factors driving the scenario. It also enables intelligence agencies to target certain important but poorly understood factors for greater attention.

Plausibility also requires complexity: surprises, perverse or unintended effects of policy, drivers interacting in unpredicted ways, are fair game. Beware "internal consistency," scenarios as straight-line extrapolations into the future, all factors lining up to support the hypothetical endpoint, imparting a sense of inevitability. Also avoid futures that describe unmitigated disaster versus peace and prosperity. Although we should be able to evaluate on a net basis which scenario is most favorable to our interests, all should pose challenges, and all should afford opportunities. Cartoonish scenarios describing extremes of best case, worse case, and muddling through are implausible, and if accepted as a basis for thinking about policy limit choices and ignore trade-offs, which defeats the very purpose of what we're trying to accomplish. Credible future scenarios are as contingent and contradictory as the present. They are credible only to the extent that they explain why some dogs barked, and others didn't.

While offering plausible narratives, the alternate scenarios must be *distinctively* and consequentially different from each other. Again, if we keep the value proposition in mind (insight, not an inevitably false certainty about the future), it becomes obvious that alternate scenarios that converge around some "most likely," and extrapolative future have lost their value from the outset. It's easy to extrapolate plausible futures, but loose the heuristic and policy value in the process—and also miss the inevitable surprises. Indeed, by going through the motions without getting the underlying purpose right, such exercises can generate overconfidence and increase the chance of surprise.

The required distinctiveness is a difficult but essential challenge to plausibility. Multiple scenarios that are both plausible and distinctive require mind-stretching exercises and persuasive exposition. They require risk-taking, a suspension of disbelief as the scenarios are constructed, and effective salesmanship about plausibility and policy implications. If there's an error to be made here, it's on the side of distinctiveness, a broadened definition of plausibility, a wide "scenario space." If the job is not to predict, but to open minds, and if participants are reminded of the costs of too narrow an understanding of plausibility, scenarios can deliver both insight and alertness, and improve the management of uncertainty.

The importance of distinctiveness does not require that each scenario be different from the others in every respect. They all begin in the present, and diverge in important ways as the narrative unfolds, sometimes slowly as drivers behave and interact differently, sometimes suddenly in response to wild cards. They thus represent distinctive directions of change that produce clearly differentiated outcomes.

These distinctive scenarios should be *comparable*. By deploying the same set of drivers across the scenarios, but varying their behavior, and by treating causation explicitly, scenario builders and readers can easily discern the causal differences and deflection points that explain how each future evolved as depicted. Factors that produce outcomes deemed positive by policymakers can be encouraged; drivers with negative effects can be targeted for prevention or mitigation. Policy choices can be revealed, as the costs and benefits of alternate actions vary across the scenarios (see the policy section of the Syria scenarios).

*Utility* for policymakers is, of course, the ultimate goal of alternate scenarios. The choice of scenario concepts to flesh out should be governed by this criterion, which in turn requires that the group understand which of the vast number of potential scenarios creates the most useful counterpoint to the "official future." Each future history must explicitly treat US policy itself as a driver (this is the fundamental difference between how to design a scenario exercise for the US policy community vs. one for a corporation), sometimes—depending on the scenario—in a positive light, sometimes not. Some of these policy effects will be deliberate, some inadvertent. Each scenario should prompt the question, are these future conditions a net benefit to our interests, or do they damage them. Across the scenarios, policymakers should ask, How can I encourage scenarios that overall are conducive to our interests and prevent, mitigate, or manage the less favorable? Because each scenario, to be plausible and useful, will describe complex combinations of risks and opportunities, each will pose challenges to current and future policy. The most plausible case for a peaceful resolution of the Syria conflict, for example, is self-evidently the "best" of the three Syria scenarios, but if pursued single-mindedly, and without enhanced US leverage (i.e., without difficult policy choices), can become the enemy of the good enough, namely containment of the conflict inside Syria (an opportunity that has been lost), and the best strategy for avoiding the worst-case, regional spillover (which is what we have now).

Clearly, utility requires honesty, self-criticism, and a willingness to expose flawed assumptions and follow the logic of the scenarios toward potentially fundamental shifts in policy. The goal should be to create productive friction with the official future. Robert Lempert, in "Can Scenarios Help Policymakers Be Both Bold and Careful?" argues that "given the multiplicity of plausible futures, the most important scenarios to consider are those few that most affect the decision-makers' choice among alternative strategies. A bold and careful strategy should exploit opportunities while avoiding the many ways in which plans can go awry. A small set of scenarios should be chosen to highlight the key trade-offs policymakers face in designing such strategies."[22]

Scenarios can, of course, be abused, though they are not inherently more subject to misguided or dishonest use than any other analytic technique or policy argument. A well-constructed future narrative, designed to be taken seriously by policymakers and others, can lose its hypothetical quality, can in the mind of the reader become a forecast, thus reinforcing overconfidence, instead of diluting it. If not facilitated professionally, or if populated by too narrow a group of experts, scenario exercises can be too narrowly conceived, thus reinforcing a priori policy commitments. They can be deliberately manipulated to serve interests, or to lay the basis for precooked and ill-considered actions. They can be pro forma exercises, designed to reassure potential critics that budget proposals for long-term commitments have undergone due diligence. They can be sited in agencies whose mandates limit the scope of the exercise, an example being the NIC's otherwise excellent alternate futures reports that cannot attribute current challenges to the effects of previous US policy. For all these reasons, it's important to do this well.

Chapter 4

# The Scenario Construction Process

The first thing to be said about the process of scenario construction is simply to emphasize its importance. Traps abound, in the subtle and easily overlooked distinction between forecasting and structured speculation, in the value trade-offs between accuracy and utility, in the difficulty of generating policy-level support without overpromising. It's very easy to go off track in a scenario exercise, and very difficult to get back on. Too narrow a range of alternate scenario concepts, passive facilitation, unawareness of policy context and of what constitutes relevant outcomes, the wrong mix of experts—these and a multitude of other wrong turns can compromise the process and deservedly marginalize the results. Good process is the only protection from these traps, the only source of consistently valuable outcomes, and the only valid basis for asking policymakers to seriously consider the results in thinking about US strategy. In effect, the "sale" to policymakers is both usable results and a defensible process. From experience, policymakers demand both.

The process itself should be transparent to all participants and to readers of the output. The process for each scenario project should be well documented and replicable. Project leaders should be prepared to explain how the alternate scenarios emerged, how drivers behaved, how wild card events shaped the narrative, why a particular set of outcomes occurred while others did not. Scenario practitioners should spend serious time working on process, experimenting with alternate approaches, and eliciting feedback from

participants on what worked, what didn't. They should be prepared to adapt process to the particular purposes and subject matter of the project in question, without sacrificing basic principles.
They should study how others do it. They should be self-critical,
prepared to learn from past mistakes. They should be devoted to
usable results, and to advancing the process state of the art.

To illustrate how to explain to one's audience how scenarios
emerged from process, reproduced below is the "Introductory
Notes" section of our Russia 2020 alternate scenarios report, based
on a workshop conducted in February 2010.

The scenario process employed during the Russia workshop
in February adhered to the format of previous Iraq, Iran, and
China sessions. The session was designed to facilitate open dialogue among leading Russia experts. There were no assigned
roles. No papers were presented.

Participants were encouraged to consider a wide range of
plausible futures and challenge conventional wisdom. The objective of the session was not to predict the nature of a 2020 Russia,
but rather to identify three materially different but plausible scenarios to increase our understanding of factors that could impact
Russia's future path. These scenarios were chosen to encompass
a wide range of conditions, including some that are of low probability but which would be highly impactful if realized, and to
challenge conventional assumptions and preferences.

The fifteen participants represented varied expertise across
economics, domestic politics, global energy markets, ethnic and cultural studies, and foreign and security policy. The
Center for Global Affairs project team opened the session by
briefly introducing a set of "starter" scenarios as a means to provoke the discussion. Session participants had previously been
provided with a background "drivers" paper, which identified
factors likely to have the greatest impact on Russian behavior

and the greatest influence on Russian policy decision-making over the course of the next decade. Participants were encouraged to challenge the starter scenarios, which they most certainly did, and to propose alternatives.

We encouraged participants to adopt an "inside-out" approach, in recognition of the criticality of Russia's internal environment. In the West, Russia is often characterized by the nature and direction of its foreign policy. While it is essential to appreciate the central role foreign policy plays in Russian political thought, fixing the emphasis on international relations often serves to obscure the highly dynamic nature of Russia's internal circumstances. As is illustrated clearly in the scenarios included in this report, Russia's future will be shaped primarily by internal forces. In fact, Russia's behavior on the global stage is a reflection of its complex internal environment—its intricate domestic politics, the demographic risks it faces, the myriad of regional and ethnic conflicts and contradictions it faces, and the geographic breadth and scope of the Russian Federation, spanning eleven times zones, and dozens of cultures, religions, and individual dialects. While Russia is a major player in the global economy and is highly exposed to changes in the global energy markets, twenty-first-century Russia is a reflection of its domestic struggle to escape from its Soviet legacies. Given this inside-out approach, participants were able to avoid deterministic conclusions about which of Russia's futures are beneficial or detrimental to US policy.

A wide range of topics and issues was discussed during the process of selecting and framing out the final list of scenarios. Initially, the focus was on the long-term impact of the global recession. Participants debated whether the global downturn would serve as a catalyst for change, propelling efforts to diversify and modernize the Russian economy, or whether the eventual economic recovery would only reinforce the current oligopolistic nature of Russia's domestic commercial environment. A number of experts questioned whether economic liberalization is possible without deep changes to Russia's authoritarian political structure. Many were skeptical that a vibrant economy and competitive marketplace are possible under authoritarian rule. Several participants cautioned that it would be very difficult for Russia to follow China's current "economic pluralism / political monopoly" model, contrasting Russian and Chinese economic circumstances and endowments. Others thought that it is possible for Russia to expand its economy without modifying the current governance model, given its vast energy resources. This was one of many instances where the experts' views and opinions diverged.

There was support for the argument that Russia's economy would continue to be highly centralized and tightly controlled from the top as long as Russia remained principally a hydrocarbon export nation. Proponents of this view pointed to the Russian government's continuing support of large state-owned enterprises, which would continue to constrain innovation and free-market competition. They questioned whether Russia would be able to gain access to the investment capital and technological know-how necessary to expand and modernize the economy without first committing to wholesale reform and implementing policies that encourage innovation and risk-taking.

Participants emphasized the corrosive effects of corruption, especially in Russia's outlying regions. Political authority was shifted from the regions to the central government in the mid-2000s in an effort to staunch the spread of official corruption. Experts debated whether political liberalization and the decentralization of political authority would ultimately lead to a loss of control and unleash new ethnic tensions in these regions, possibly leading to economic or political fragmentation.

The focus then shifted to the likelihood of dramatic change in Russia over the next decade. There are few examples of bottom-up or organic change in Russian history. Several participants noted that most instances of structural change resulted from unilateral decisions made by Russian rulers or as a result of elite power struggles. There was a lengthy discussion about how changes could occur in the context of Russia's current political structure. Most were not persuaded that organic or bottom-up change would be possible. Participants debated the importance of popular support and questioned whether the Russian public would continue to support the current political structure without substantial improvement in Russia's economic circumstances. Conversely, many of the participants challenged the notion that a major economic or political shock (e.g., external conflict, wave of terrorism, economic calamity) would lead to a rejection of the current system. They argued that Russian citizens historically rally to the government in times of crisis.

After this lively two-hour discussion, participants were able to identify three alternative scenarios that reflect the highly variable nature of Russia's political and economic environment. As has been the case in all the previous scenario sessions, participants were asked to suspend disbelief, set aside probabilities, and build the most persuasive case for each scenario. All three narratives begin in the present, with the same economic and political circumstances, but they begin to diverge significantly as a result of how policymakers in Moscow react to the unfolding events and circumstances;

- *Working Authoritarianism*: After a prolonged period of economic stagnation, Russia is forced to dramatically shift its policies to jump-start growth. It enacts a draconian austerity program that includes wage ceilings and the elimination of the household electricity subsidies. It also establishes strategic partnerships with China, Germany, and South Korea, which enable Russia to gain access to much-needed capital, technology, and know-how without being subject to "conditionality."

Thus, Russia is able to energize and diversify its economy without political liberalization or reform.

- *Bottom-up Liberalization*: The struggle for dominance between reformist and conservative elements in the Moscow power structure results in a stalemate, reducing the government's ability to act decisively to address economic drift. Fueled by the dynamism of a new generation of entrepreneurs and seed money from Moscow, new industries and commercial enterprises emerge in a number of Russia's regions. This symbolizes a Russian economic rebirth and paves the way for new political forces to gain traction.

- *Degeneration*: Deep into the new decade, the Russian government remains unable to solve the country's deep economic problems. But Moscow is unwilling (or unable) to tolerate alternative solutions that might weaken the regime's grip on political power. The country continues to stagnate, forcing the regions to "fend for themselves." This further reduces the Russian state's capacity to govern and leads to political and geographic fragmentation.

The final part of the session was devoted to a discussion of what we learned, or had confirmed, about possible change in Russia and in US-Russian relations. We then opened the discussion to other faculty, students, and observers for questions and comments.

The scenarios delineated in this publication are the product of a rich discussion encompassing a full range of topics and perspectives. They are not intended to be mutually exclusive. Each scenario, however, represents a dominant tendency with distinctive implications for Russia, its neighbors, partners, and rivals. They are not predictive in nature. Rather, each is a plausible picture intended to demonstrate the nonlinear nature of change in Russia. The hope is that by employing this alternative scenario format and processes, we have been able to reveal challenges and opportunities for US policy that might not become apparent in more traditional, current policy-driven debates.

## Design of Scenario Process

There are important up-front design choices to be made about how to organize any alternate scenario project. These choices turn on the strategic interests of the client or consumer, the subject matter, and the time frame. It is critical that these variables be assessed at the outset, and the approach molded to overall purpose and context, if the full value of the exercise is to be achieved. There is no single cookbook for scenario construction. What must not vary is the commitment to plausible, distinctive, and consequential (useful) results, but the road map is highly variable.

Broadly, two approaches exist for scenario generation. "Bottom up" approaches begin by identifying drivers—factors that are important in shaping the future of the country or issue, their dynamic interaction, and their variability over the scenario time frame. Alternate scenario ideas emerge as the drivers are combined and varied, often with the aid of a matrix that can help in identifying plausible and consequential clusters of drivers. The group can then look across the most promising driver intersections, and choose a limited number—usually three or four—for subsequent development into scenarios. Such a bottom-up approach places the emphasis on broad questions of change in a given country, issue, or the global system, and may be occasioned by prudential considerations (the risk associated with current policy is increasing, and a rethink of assumptions is called for), a new administration interested in re-examining previous assumptions, a surge in capacity that enables more ambitious aims, or a surprising and consequential event that invalidates current policy. The scope of analysis is broad, and policy implications emerge from the exercise, more than shaping it.

It is often asserted by scenario builders that scenario exercises should be organized around a policy issue or choice facing the client, that such specificity allows a narrowing of scope while maximizing relevance and utility. In effect, they advocate a top-down process, defined and scoped by some current or imminent policy decision (should we arm the Syrian opposition?) or broad strategic choice (how can we mitigate potential challenges to the pivot toward Asia?). Such top-down approaches link directly into the policy process, are targeted at shorter-term challenges to existing policy or strategy and seek to add value to policy debates without overthrowing existing paradigms.

In our scenario project funded by Carnegie, the initial objective was to target "pivotal countries" and to assess longer-term (to 2020) future possibilities, without reference to current US policies or interests. We argued in defense of this approach that big events in such pivotal states would be consequential regardless of current policies, that the policy value and market for such output would be strong because in fact some of the scenarios would be policy forcing. To organize around a specific current policy priority or impending choice could also deliver value, but could conservatively bias the results toward those deemed relevant to contemporary priorities, when in fact US strategy has fluctuated greatly in response to unanticipated events and is certain to continue doing so. In effect, the scope of any project organized around current choices as defined by policymakers can become in-box driven, subject to the very biases alternate scenarios are designed to overcome. Better, we thought, to demonstrate the current policy implications of unfamiliar future possibilities than to guarantee "relevance" at the risk of missing the game changers.

In the first, four-country project for Carnegie (China, Russia, Ukraine, Turkey), and in a subsequent study of Pakistan we followed primarily a bottom-up approach, building toward scenario ideas from the underlying drivers of change and their interaction, and from potential wild card events, imagining the multitude of combinations that might result. I say primarily because one of the criteria for choosing three among several possible scenarios for detailed treatment was utility for assessing the robustness of current policy. The bottom-up identification of potential futures maintained the necessarily broad scope. The utility criterion applied to selecting scenarios for deeper consideration assured relevance. The details on this approach are addressed later. The point here is that the longer the time frame and the greater the potential variability of the country (or issue), the greater the value of bottom-up approaches and the greater the risk of current policy-driven scenarios that encourage a false sense of confidence and miss future policy-forcing events.

When asked by Carnegie in 2013 to extend the project to Syria, we rethought this approach. Syria was by this time in the midst of civil war, its future almost entirely dependent on the outcome of that conflict. The logic of the conflict trumped long-term drivers that, under more stable circumstances, would collectively have shaped the country's future—quality of governance, conflicting sectarian and national identities, economic growth and employment, demographic trends, regional rivalries—the type of forces we considered in other "pivotal countries." We thus shortened the time frame, from 2020 to 2018, set aside the drivers of Syria's very long-term, postconflict future (if it has one) and focused on alternate possibilities for the war's outcome, from peaceful resolution (through partition or power sharing) to regime or opposition victory, to contained stalemate, to escalation, fragmentation, and regional spillover. These candidate scenario concepts, from which the panel chose three for detailed treatment, emerged quickly from the literature and ongoing public debate about the future of Syria. All were directly relevant to US policy: downsides challenged US assumptions about opposition victory or US ability to contain the conflict without a larger investment; upsides asked whether getting to better outcomes also demanded greater US leverage. This more "top down" approach to selecting useful scenarios worked well in circumstances that did not permit us to assume that Syria had a future: looking at macro-drivers would have been pointless.

## Selecting Participants for Scenario Workshops

Whether the chosen design is primarily top down or bottom up, the scenario construction process should be interactive, a structured "brainstorm" among experts and policymakers of diverse professional backgrounds, nationalities, and skill sets. The spontaneous interaction among such individuals brings a wealth of expertise into the project, facilitates cross-disciplinary insight, broadens the network of skilled scenario builders and proponents, and enhances the legitimacy of the final product. The diversity of the group reflects the fact that major, unanticipated change is often a product of trends and events from different domains of activity intersecting in unpredictable ways. A dialogue limited to international relations experts and policymakers may be circumscribed by traditional IR theory, which places many forces for change outside the dominant models (technology, ideational change, economic variables, demographics, climate change). A dialogue among Americans of diverse skill sets may fall short on local knowledge (in our Syria project, Syrian nationals consistently rated Assad's chances for a comeback as greater than did Americans), particularly for countries historically peripheral to US interests. A special contemporary liability of our vast power is indifference toward the views and interests of others, often manifested in US-centric assumptions of how others think and behave ("mirror imaging"). A group with too many country-specific experts may fail to incorporate regional or global factors, or stick to deeply grooved conventional wisdom about that country (Pakistan will always muddle through; the Saudi monarchy will always find a way to survive; the Soviet state is impervious to change). Too many of the usual suspects may militate against risk-taking and produce too narrow and predictable a conversation. A panel composed of "futurists" may leap to startling, out-of-the-box insights, but may lack the expertise to ground the scenarios in present circumstances, to produce plausible narratives or identify policy implications.

This is not a meeting of stakeholders. Again, the value added is insight into policy-relevant future trends and events, especially those not on the policy radar. It is not to achieve consensus among all interests who are considered important in a given policy domain. I have often had to resist pressure from sponsors that a particular research center or set of NGOs should be "represented." Such meetings may have a point, but invariably suffer from stakeholder posturing, well-rehearsed positions, and resistance to considering undesirable scenarios. The only stakeholder that counts here is the policy community. Otherwise, criteria important in choosing participants are relevant expertise and imagination.

Experts should be chosen carefully. The experience gained through our pivotal countries project suggests that country experts often underestimate the degree of future variability in the domestic politics of seemingly stable states. That is to say, they tend to dismiss signs of current tension, minimize the capacity and ambition of regime opponents, and exaggerate the resilience of existing authority (they were correct on Assad, but in part because American restraint gave the better-supplied regime a big edge). This is the case across the Middle East, as it was with the Soviet Union. The point is not that experts fail to predict revolutionary events, but that they often resist a serious consideration of such possibilities, thus artificially narrowing the conversation, depriving policymakers of the combination of expertise and open mindedness essential to early warning, and limiting the opportunity to test policies in the presence of such changes. This is all the more remarkable given the centrality of the Middle East to US interests, the sheer amount of attention, in government and the academy, devoted to watching it, and the potential costs of being unprepared for big changes.

Expertise is of course essential for effective, credible scenario exercises, and the best in their respective fields should be recruited. The process itself, if conducted with skill, will push many of them beyond their comfort zones. Others will instantly understand the particular value proposition, and will welcome the opportunity to think outside their disciplinary boxes. Some will be carried along as the conversation gets interesting. Others will dig in their heels, asserting that some seemingly plausible scenario or event "couldn't happen." In my experience, a mix of country and regional experts from academic institutions, governments, and think tanks, observers of overall global developments, former and current policymakers, and public intellectuals is optimum for such workshops. It's also helpful if a few have already participated in such exercises and are familiar with the "willing suspension of disbelief" essential to moving toward useful results. The downside here is that such individuals may think they have a better process than yours and attempt to revise the approach during the meeting. A full explanation of process at the outset of the workshop, and strong facilitation to preempt process debates, is necessary.

I discussed the fox versus hedgehog distinction in chapter 3. Because this is an interactive group process and not individual expert polling, the correct mix of expertise and cognitive styles is more important than individual forecasting "accuracy." As (if) Tetlock confirms the consistent superiority of his cadre of "superforecasters" and their success in predicting change as well as continuity, one would want to make it a point to engage them in scenario-building workshops. The proposition here, however, is that groups of experts with skill sets matched to the diverse drivers of change, and good process, will produce plausible, distinctive, and useful scenarios superior to brilliant individuals who are measurably better forecasters. Both hedgehogs and foxes should be welcomed to these exercises, the former for their self-assured, parsimonious constructions of the future, the latter for their open, inquiring style and devotion to debate. Of course, foxes and hedgehogs don't usually identify as such, so selecting participants based on cognitive style is sometimes guesswork. But experts do acquire reputations over time, not just for their expertise but for their flexibility and openness to challenge, and these inclinations are discoverable as one goes about the process of selecting participants.

The particular skills necessary for a successful workshop will, of course, depend on the country/region/issue, and the drivers considered critical in shaping the future. In order to achieve the necessary cross-disciplinary conversation, the fifteen or so participants should collectively cover the drivers, have country and region expertise, take a global view of politics and economics, and have experience (ideally currently) in the foreign policy process. Some of the country experts should be nationals. As the preworkshop research on drivers produces a sharper sense of the specific issues facing a given country (water scarcity in Pakistan, national identity in Ukraine, etc.) experts covering these issues can be identified. With one or two slots reserved for very smart generalists who understand scenario thinking but have no portfolio, sticking to fifteen participants isn't easy, but more than that number begins to degrade the quality of conversation.

Identifying specific participants to invite can be both a top-down and a bottom-up process. I start with my own network of experts in the academy, think tanks, and government, asking those I trust, including former participants, to suggest who would provide the particular expertise we need (matched to the drivers) and would take easily and enthusiastically to the alternate scenario format. For both issues and countries, there are well-known academic centers of excellence. Bottom-up selections come from the drivers research (see below) and can produce brilliant, often young but underrecognized experts that the informal process would be unlikely to identify. A group assembled in this manner will have the necessary scope and diversity of skills and points of view to get the project off on the right track. Such a group also enhances the learning experience for the participants far more than a group composed of the usual suspects, and is itself a selling point in attracting the people you want.

## Preworkshop Research on Drivers of Change

The identification and analysis of drivers of change imparts structure to the exercise, taking it well beyond spontaneous brainstorming among smart people. Drivers provide a common vocabulary, a degree of rigor, transparency, and replicability of results. They facilitate a comprehensive discussion of the range of factors shaping the future and permit explanations of how the alternate futures emerged from the present. In a bottom-up process, the interaction of drivers allows the systematic identification of alternate scenario ideas from which the group can chose a few for detailed treatment. Their variability explains how and why the alternate scenarios diverge. The analysis of drivers helps to organize the preworkshop research, identifies experts in relevant fields for recruiting into the workshop, and generates read-aheads for participants that help prepare them and project staff for the workshop.

Countries, regions, and issues are, of course, more than the sum of a few key causal factors. Drivers are in themselves oversimplifications of complex realities. History is full of contingencies, and the future is indeterminate. Drivers allow us to set up a systematic process, more likely than haphazardly organized conversations to include most major variables and important interactions, embrace alternate points of view, and produce plausible results. But in providing structure, they do not determine outcomes. They help to launch the conversation, but the spontaneous interaction of participants will always take the group well beyond the scope defined by drivers.

In structuring research on drivers, we try to identify a few key factors that singly, or more often in combination, are considered by experts to shape the present and future of the country at issue. These include economic variables (growth and employment, inflation, income distribution, resource endowments), political variables (who governs, political succession, rule of law), quality of governance (legitimacy, leadership, corruption, stability), demographics (size and growth rate, age, ethnic, racial composition, migration), identity (national, regional, ethnic, sectarian), civil society (strength, composition, impact on government), external interactions (within the region, policies of outside powers, including the United States). The role of US policy and power is always in play, either its presence or its absence. Such drivers tend to be standard across countries, though their relative importance varies greatly.

The drivers paper, done by the scenario staff and sent to participants well in advance of the workshop, follows a common template: why is the driver considered important? What is the relative importance of the drivers for that particular country? What is the plausible range of variability in the driver over the time period of the scenarios (demographic variability is usually quite low, economic growth has varied historically for the country from −3 to +5 percent, etc.)? What potential driver interactions might produce big surprises over the scenario period? What potential scenarios and wild cards emerge from the analysis of drivers?

Reproduced below are two examples of drivers papers, for China and Ukraine.[1]

---

## DRIVERS PAPER

### *China 2020*

The Chinese political system of the past twenty-five years has rested on an implicit understanding between the Chinese Communist Party (CCP) and the people of China, whereby the government delivers economic development and ever higher levels of prosperity in return for popular acceptance of the CCP's monopoly of power. Will this system of government continue to deliver its end of the bargain? The Chinese leaders must now address some significant problems if they are to repeat the economic successes of recent decades:

- The negative consequences of a flawed economic model, specifically, high levels of corruption, decreasing productivity levels, unprofitable state-owned enterprises (SOEs), and state-owned banks with high levels of bad loans

- The growing levels of inequality and ethnic strife and significant environmental degradation caused by rapid industrialization and high growth rates
- The growing pressures of globalization in terms of price competition, higher quality standards, and global expectations for further trade/currency liberalization
- A high dependence on export trade and investment as sources of economic growth, both of which are much less stable and sustainable than domestic consumption

The degree to which the Chinese political system will address current challenges is widely debated. The recent Center for Strategic and International Studies publication *China's Rise: Challenges and Opportunities* presents three possible scenarios that might unfold:

- Minxin Pei's vision of a China "trapped in transition" whereby China remains indefinitely under authoritarian rule, plagued by crony capitalism and unable to get to the next stage of economic development[2]
- A more hopeful vision from Bruce Gilley, who foresees the better-educated and more affluent emerging middle class becoming a major force for change, leading to significant reforms within the CCP or possibly even a change in regime by 2025[3]
- Randal Peerenboom's hypothesis that China will follow the East Asian development model of countries such as Singapore, where its economic growth will be managed under a system of softening authoritarianism to eventually emerge as a "democracy with Chinese characteristics"[4]

A key determinant of the trajectory China ultimately follows will be how it emerges from the current global economic downturn.

Pivotal to this will be the CCP's ability to rebalance the Chinese economy on the production side toward a greater emphasis on services rather than industry and on the demand side away from export trade and investment toward domestic consumption.[5] This has officially been part of the government strategy since December 2004, and a successful execution would help address many of the current challenges:[6]

- Growth would be less capital intensive, enabling China to grow with a lower level of savings and become less intensive in its use of energy and other natural resources, leading to lower carbon emissions and less environmental damage.
- Inequality would be reduced or at least its growth slowed, as average personal consumption levels rise and the reduced focus on export-driven growth leads to a greater economic balance between coastal and inland areas. World Bank research suggests that China's heavier focus on industry, as opposed to services, has left China less able to absorb its excess agricultural labor. The creation of more labor-intensive urban jobs, particularly in services, would help China improve total factor productivity by enabling more rural dwellers to find work in urban centers, reducing rural poverty levels and the urban-rural income divide.[7]
- The rebalancing would lead to a reduction in China's trade surplus, returning greater stability to the international system and reducing the likelihood of protectionist policies being pursued by the Americans and the Europeans.

In the current global environment, we must ask some key questions: is this goal of rebalancing the Chinese economy achievable? Has the CCP demonstrated a commitment to this policy, or will the party in fact be forced down this path? What are the primary levers to such a transition? Will this strategy inevitably lead to lower average growth rates? How might the various internal and external stakeholders in the Chinese system respond to these changes? What might the implications be for the United States of diminishing interdependencies with China?

## Economic Growth

In late 2008, the Chinese government took decisive action to get the economy back on track. China announced a stimulus package of RMB 4,000 billion (US$585 billion) and RMB 7,370 billion of new bank loans. The loan component of the plan is believed to have created more credit than any other economy has created since the end of World War II.[8] This level of investment is credited with creating the V-shaped recovery we are now seeing in China, which marginally overtook Germany in the first half of 2009 as the world's largest exporter.[9] Chinese GDP growth in Q2 2009 returned to 7.9 percent and it is now projected that China may achieve an average growth rate of 8 percent or more in 2009.[10] This is still below the double-digit growth of recent years.

While the initial impact of the stimulus is broadly welcomed, experts highlight flaws in its structure, such as the possibility that new asset bubbles may be created in equities and property, that more bad loans have been made, that most of the investment has gone into energy-intensive infrastructure projects (namely roads, rail, and airport projects) that are likely to create overcapacity in certain areas, that these large infrastructure projects have generally created replacement jobs for some of the growing numbers of unemployed without expanding overall employment opportunities or leading to increased domestic consumption, and that the stimulus has mostly gone to large SOEs where waste and corruption are rampant, while ignoring SMEs (small-medium enterprises) that provide up to 75 percent of jobs in urban areas.[11]

It is thought that this heavy focus on fixed-asset investment may generate only short-term benefits and that this could transform China's economic recovery from a V- to a W-shaped recovery. It is broadly agreed that the Chinese must look to domestic consumption as a future source of growth, which presents many challenges. The Chinese population typically spends less on personal consumption than their Western counterparts, primarily because of the absence of an adequate social safety net. While earning less than Western workers, Chinese workers have to save to cover the medical and educational needs of their families. If the government were to help cover those costs, people might be encouraged to spend more and save less.[12]

Nicholas Lardy suggests that there are several levers available to the Chinese government to rebalance demand, specifically through changes in the following four policy areas:[13]

- *Fiscal policy*: The government should reduce income tax rates, increase minimum wages, spend more on the provision of healthcare and educational services, and possibly fund incremental expenditure through higher dividend taxes on SOEs.
- *Financial reform*: The current low interest rates paid on consumer savings accounts are not keeping up with inflation and are subsidizing corporate borrowers, who continue to drive excessive levels of investment in energy-intensive industry. Higher consumer interest rates would increase disposable income levels, encourage more consumer spending, and reduce the availability of cheap capital for investment.
- *Exchange rate policy*: Liberalizing the exchange rate policy and allowing an appreciation of the renminbi would lead to a reduction in exports and an increase in imports, reducing the weight of exports in economic demand.

- *Price reform*: Ensuring full-cost charging for energy, water, utilities, land, and environmental impact would discourage investment in energy-intensive, environmentally damaging industries and encourage more investment in services.

Jianwu He and Louis Kuijs suggest some additional areas for policy change:[14]

- *Service sector*: The opening up of the service sector to private and foreign investors and the creation of a legal framework that would encourage growth in this area
- *Houku system*: Changes to this system that limits migration to urban areas, discriminates against migrant workers, limits the transfer of labor and social benefits, and enforces restrictive land tenure policies

Changes in the above areas would impact a broad base of stakeholders, positively in the cases of rural dwellers and migrant workers and negatively in the cases of entrepreneurs, who would pay more for access to capital and SOE managers whose profits would be reduced.

Which options would be the most attractive to the CCP? Which options are party leaders likely to pursue and within what time frames? What are the likely impacts going to be and over what time frame?

## Energy Supply and Demand

China's energy consumption has grown rapidly over the past three decades. China's ability to get continued, secure access to enough energy to fuel its economy in a highly competitive global energy market is a major government concern. Primary energy consumption quadrupled between 1980 and 2007, and aggregate demand is projected to double again by 2030 to 3,128 million tons of oil equivalent (Mtoe).[15] China is expected to overtake the United States as the largest energy consumer within two years. China's dependence on imported oil is projected to increase to 50 percent by 2010 and to 60 percent by 2020. As early as 2015, it is predicted that 70 percent of China's oil imports will come from the Middle East through the Strait of Malacca. Coal-fired power, which emits high levels of carbon dioxide, provides 66 percent of China's energy, 76 percent of electricity generation, and 70 percent of all energy consumed.[16] To meet this growing energy demand, China is building two new coal-fired power stations every week.

Given the economic imperative of accessing affordable energy, the security implications of a high dependence on Middle Eastern oil and the environmental consequences of a heavy dependence on fossil fuels, in particular coal, the Chinese are working hard to diversify their energy sources. This includes increased investment in renewable energy technologies and controversial strategies exemplified by their relationship with the Sudanese government. The Chinese are investing in many developing countries and providing financial aid to those nations without any conditions in return for access to natural resources. This strategy is a source of increasing conflict between China and Western powers, particularly where human rights are being violated.

## Environmental Stresses

China is now the largest emitter of carbon dioxide in the world, and the Chinese people are already seeing the impact of climate change in the expanding deserts of the north and in the increasingly violent storms hitting the south. The melting of the Himalayan glaciers over the coming decades will lead to extreme water shortages as the main river arteries of China and its neighbors to the South start to dry. The melting of the polar ice caps, which could raise average sea levels by several meters, would leave coastal cities such as Shanghai completely submerged. China is already seeing some internal migration due to expanding deserts in the north but faces the prospect of mass movements of climate migrants in the future.

In June 2009, the CCP announced its intention to introduce carbon dioxide emissions targets for China's various social and economic programs. This was viewed as a signal that national carbon emissions targets might be included under the next five-year plan, which will run from 2011 to 2015.[17] In August 2009, Premier Wen Jiabao declared that climate change considerations should be built into the medium- and long-term development plans of all levels of Chinese government. At the UN Climate Change Summit on September 22, 2009, President Hu Jintao announced that China would reduce $CO_2$ emissions per unit of GDP by 2020 to a level significantly below 2005 levels and increase investments in reforestation and renewable energy sources.[18] Although he did not specify $CO_2$ reduction targets, his speech signaled that China is now engaging fully in the fight against climate change. While welcome at the global level, this new commitment will face some resistance within China. To date, one of the weaknesses of the Chinese strategy has been its inability to ensure nationwide implementation and enforcement of new climate change legislation.[19] A key question, therefore, is whether or not the central government will be able to successfully execute the ambitious targets set by President Hu.

### Inequality across China

Since 1981, China's poverty, measured by the World Bank as the number of people living on less than $1.25 per day, has dropped from around 65 percent of the population to just 10 percent in

2004. This decline represents an estimated 500 million people who have been lifted out of poverty.[20] While most Chinese households have benefited from the economic boom, there has been a significant increase in income inequality. The GINI measure in 1981 was just .31, but by 2005, it had risen to .46.[21] This widening gap can be attributed to both the market-based system, which created incentives for wealth creation, and to features of government policy, such as restrictions on the ability of rural residents to move to cities, the prohibition on the sale or mortgage of rural land, and the decentralized financial system, which leave local government responsible for the provision of health and educational services.[22]

As factories close across the major urban centers during 2009, inequalities are being exacerbated. It is estimated that up to 20 million migrant workers will find themselves unemployed in 2009, a trend that has been described as a "ticking time bomb for the Communist Party."[23] How might these migrant workers and their host communities react to continued and increasing high levels of unemployment?

### Ethnic Minorities

China is home to fifty-six officially recognized ethnic groups, of which the dominant group is the Han. According to the 2000 national census, 108.46 million Chinese people were members of an ethnic minority, 8.04 percent of the total population.[24] The remaining 92 percent consisted of the dominant Han. They too are comprised of different subgroups, including the Cantonese, the Fujianese, and the Hakka. Although the government positions the Han as a homogenous unified ethnic group, eight Han subgroups speak completely distinct languages.[25] It is among the Han subgroups that the Chinese government is thought to see a potential future threat to Chinese stability.[26]

China's Muslim population is estimated at approximately 20 million people dispersed across ten different ethnic groups, with over 70 percent living in Western China.[27] The largest group is the Hui, who are culturally Chinese and either converted to Islam or intermarried with Muslim migrants. Many other Chinese Muslims are Turkic in origin and directly linked to minority groups across Central Asia or to populations within the former Soviet republics.[28]

Officially, the Chinese government promotes equality among all Chinese people, regardless of ethnic origin.[29] In practice, the government has taken steps to reduce the power of ethnic minorities through job discrimination and by resettling Han Chinese in areas dominated by ethnic minorities (e.g., Xinjiang and Tibet). While ethnic minorities are unlikely to overthrow the Chinese system, "Cultural and linguistic cleavages could worsen in a China weakened by internal strife, an economic downturn, uneven growth, or a struggle over future political succession."[30]

Given the current challenges faced by China, are we likely to see more of the ethnic fueled violence witnessed recently in Tibet and Xinjiang?

Could we see Chinese Muslim communities forging closer relationships with Islamic communities across Central Asia?

## Demographics

The Chinese population will continue to grow over the next twenty years and is predicted to peak at 1.5 billion people in 2035.[31] The Chinese birth rate is falling and now stands at 1.7 children per woman, versus the recognized replacement level of 2.1. While this should be welcome news, it is problematic in the long term when coupled with a rapidly aging population. Today, 145 million Chinese, 11 percent of the population, are aged sixty or more. By 2050, this number will rise 33 percent. China is also experiencing a growing gender imbalance. The gender ratio now stands at 117 males per 100 females, and the government fears that the growing proportion of single men in Chinese society may lead to major social problems in years to come. It is clear that demographic trends will have a significant impact on China over the coming decades. It is estimated that rural to urban migration will continue and that an additional 10 million people will leave rural areas each year to move to the cities. By 2050, it is estimated that 70 percent will live in cities, compared to 40 percent today and 20 percent in 1978. This will significantly strain the infrastructure of Chinese cities and towns and natural resources, in particularly water. China is estimated to hold just 7 percent of global freshwater sources, despite being home to 20 percent of the global population. Climate change will further exacerbate this problem. Urbanization will also accelerate the growth in the Chinese middle class, whose values and priorities diverge from their parents' generation.

A much smaller workforce will be responsible for supporting a much greater proportion of Chinese society, which holds major implications for average incomes and economic growth. The shrinking of the Chinese workforce will also damage China's competiveness in decades to come, as India's labor force eventually surpasses China's in size.

The pressures and frustrations that these changing demographics will create, coupled with a slow economic recovery and other either perceived or real grievances against the government, could lead to significant social unrest. What policy changes might the CCP introduce over the coming ten years to address some of these demographic challenges and to mitigate their impact?

### Civil Society and the Internet

The number of NGOs in China has grown rapidly to an estimated 354,000[32] by the end of 2006. These organizations increasingly fill the gap between the range of services provided by the government and those required by the Chinese public.[33] The sector is heavily regulated, and all NGOs must secure the support of a government agency to operate. As a result, a vast number of unofficial or illegal NGOs exist across China, estimated by the World Bank to number more than one million.[34] Many of these unofficial NGOs operate in areas that challenge government policy, such as human rights or religious freedoms. Over the past decade, the number of public protests in China has increased from an estimated 8,700 in 1993 to 74,000 in 2004. The average number of people taking part in these protests has also increased from an average of eight in 1993 to fifty in 2004.[35]

The Internet has become a great enabler of civil society in China. The number of Internet users had increased to 338 million as of June 30, 2009, of which 320 million had broadband access.[36] The CCP polices the Internet and tries to censor all materials published. However, emerging blogs have proven elusive for the Chinese Internet police, and blogging has become a very powerful tool for the Chinese to express their criticisms of the Chinese government. By 2007, it was estimated that almost 47 million Chinese bloggers maintained around 73,000 blogs, with 17 million bloggers updating their blogs at least monthly.[37] Although the government does systematically close down blogs that publish unauthorized material, once a blog is published, it is very difficult to remove its contents entirely from the Web. The Chinese government focuses its attentions on blogs it finds "criticizing the state or state policy directly, those advocating mass political action, or those airing views that openly conflict with party ideology."[38] However, bloggers have developed clever techniques to avoid detection, such as using codes decipherable by readers that elude their censors.[39] Going forward, advancements in Internet technologies and in the sophistication of Internet users will make it increasingly difficult for the CCP to control the Web, particularly its use by the better-educated and affluent Chinese middle class.

## Political Change

In recent years, intellectuals with very divergent views on how to address the challenges faced by China have been brought into the internal Chinese political debate, expanding the range of options under consideration by the Chinese leadership. "The United States must carefully observe these debates to improve its understanding of both how and why the Chinese leadership will choose to address the many economic, political and diplomatic challenges it faces."[40]

In October 2007, President Hu appointed two successors to the Politburo Standing Committee, Xi Jinping, an elitist (or princeling) who is expected to succeed Hu as president, and Li Keqiang, a populist who is expected to succeed Wen as premier. These two individuals come from very different backgrounds with different value sets: Xi a strong supporter of the coastal entrepreneurial classes, favoring further international integration and growth in international trade, while Keqiang is a more traditional CCP member with strengths in propaganda and organizational management and a tendency to focus on relatively populist policies relating to agricultural reform and social services. "In elevating both Xi and Li in 2007, Hu signaled the importance of the different constituencies each represents and the belief that only consensus-building will successfully forestall serious political upheaval in the so-called fifth generation of leaders, of which Xi and Li are members." [41] Their primary shared objective when they come into power in 2012 will be to transform the structure of the Chinese economy from the current export- and investment-driven model to one that is driven by domestic consumption.

Given their divergent interests, how will Xi and Li work together? Which set of values will be more to the fore? How might politics change if the economy fails to recover by 2012?

## Actors outside China

As the most populous nation on earth and a global economic and military powerhouse, China and its leadership will be strongly influenced by external drivers over the course of the next ten years, including

- The speed with which consumer confidence in the United States and Europe recovers and demand for Chinese exports starts to grow again
- The degree to which US and EU leadership decide to pursue protectionist policies against China
- The extent to which the United States and other Western powers pressure China into changing its exchange rate policy
- The likely power transition in North Korea: who will be the successor to Kim Jung-il; is there a risk of state failure; how might the role of the United States change on the Korean peninsula?
- The evolution of China's relationship with Japan and its new administration's engage strategy
- Continued competition and tension with India: will India pursue a policy of constructive engagement with China, or might it pursue a more aggressive competitive strategy?
- External players' perception of China's intensions toward Taiwan
- The world's perception of the rise of China: will it be viewed through a liberal lens or through a realist lens; if the realist narrative were to take over, how might the Chinese adjust its strategy to the outside world?
- other regional powers' response to China's growing economic strength and regional dominance
- changes to the US administration in 2012 or 2016: what might they mean for United States–China relations?

## Ukraine 2020

Ukraine is a young country with a long history. The name "Ukraine" means borderland, which is what the region has been for thousands of years: between the plains and forests; between Roman Catholicism, Eastern Orthodoxy, and Islam; between the Russian empire and its Western neighbors; between communism and capitalism; between the Soviet Union and its European satellites; and, most recently, between Russia and the European Union.[42]

Given this history, Ukraine's emergence as a sovereign nation-state and coherent polity has been surprising. In the last decade, it has continued to defy predictions, at times abruptly changing its foreign policy goals, undertaking a dramatic push for democracy in 2004, descending into unprecedented political gridlock thereafter, and suffering a 15 percent decline in real GDP in response to the global financial crisis. What unexpected changes could occur in Ukraine in the coming decade? How will its vulnerable geopolitical setting in an increasingly volatile international environment affect those changes? What will Ukraine look like in 2020 as a result?

Ukraine's future course matters as much for its neighbors as for its own citizens. The country's 46 million inhabitants, large economy, vital role in natural gas transport to Europe, and strategically important location play a pivotal role in regional stability and prosperity. Ukraine will inevitably remain a borderland, and it remains common to question whether it will ultimately turn east or west or find a balance between the two. Equally significant—and perhaps relatedly—will Ukraine find a path through its internal frictions, institutional weaknesses, and economic struggles?

### Drivers of Change

The scenarios that are plausible for Ukraine in 2020 are rooted in trends evident today. Over the coming decade, political, economic, and social forces will evolve and interact with one another, such that in 2020, Ukraine could look substantially different than it does today. This paper identifies six areas in which trends currently evident could vary significantly in the coming decade, driving change in Ukraine: political dynamics, questions of identity, demography, the economy, the energy sector, and the country's foreign policy orientation. How will these drivers of change evolve in the coming decade? What could enable them or prevent them from trending in a particular direction?

## Political Dynamics: Institutional Weaknesses and Power Politics

Ukraine achieved independence "as much by accident as design."[43] The country chose to pursue democratization on the European model, but in the absence of a genuine revolution, the transition left Soviet-era governance structures essentially intact and Soviet-era elites in power.[44] The 2004 "Orange Revolution" raised expectations, but Ukraine continues to face the twin challenges of developing effective institutions and managing power politics. To date, these challenges remain unmet, while their consequences—constant elite infighting, periodic political crises, and poor-quality governance—grow ever more visible.

*Institutional development*: Ukraine's state-building efforts since 1991 have resulted in a system that can be described as "democratic" and "free."[45] However, institutional development has proceeded slowly and unevenly. Ukraine has yet to develop clear, generally accepted "rules of the game" to govern its political system. The 1996 constitution has been amended numerous times, but the result has not been a more coherent governing framework. Instead, after each election, incoming political actors manipulate the rules to their own advantage.[46] In this environment, de facto power has become central to the system, rather than the rule of law. As a result, government bodies at all levels consistently fail to meet public expectations, manage political conflict, and provide the necessary support for a market economy. Ukraine has experienced particular difficulty in establishing a system of checks and balances that restrains executive power but still allows the government to function. The 1996 constitution contained a number of ambiguities regarding the division of powers between the president and parliament, and consequently sparked a power struggle between the executive and legislative branches as each attempted to consolidate its sphere of authority.[47] Leonid Kuchma (president, 1994–2004) succeeded in establishing his dominance over state intuitions by leveraging administrative resources and his influence over the judiciary, the police, and the media.[48] His term ended in crisis when his attempt to engineer victory for his chosen successor, Viktor Yanukovych, sparked a popular revolt, the Orange Revolution of 2004. The resolution of the Orange Revolution brought Viktor Yushchenko (president, 2005–2010) to power, but on the condition that the constitution be amended to empower the prime minister vis-à-vis the president. Far from clarifying the roles of the executive and the legislature, the new arrangement added an additional layer of complexity: competition between the president and the prime minister for control of the executive branch, a struggle that proved crippling.[49] The election of Viktor Yanukovych in February 2010 appears to have broken the deadlock temporarily, but some observers worry that Yanukovych's efforts to consolidate his authority have the potential to threaten recent democratic gains.[50] On October 1, Ukraine's Constitutional Court overturned the 2004 law that had shifted powers to the prime minister,[51] reinforcing once again the central role of the president.

Struggles between the president and parliament have been exacerbated by problems within the parliament itself. The absence of strong, ideologically oriented parties and stable majority coalitions has hindered the functioning of parliament and prevented it from acting as an effective counterweight to presidential power.[52] Ukraine's party system is highly fragmented, reflecting the diversity of Ukrainian society, the consequences of electoral laws, and, some argue, political elites' unwillingness to compromise.[53] Changes to electoral laws in 2004 brought measured improvement: only five parties were elected to parliament in 2007, compared to twelve in 1997.[54] Nonetheless, forming majority coalitions continues to be difficult, as evidenced, for example, by the months required to complete the process after the 2006 parliamentary elections and 2007 preterm elections.[55]

*Power politics*: Power has tended to concentrate very quickly in Ukraine: "Those that have it get more and more."[56] Given that Ukraine's post-Soviet transition was engineered by Soviet-era elites—to their own economic and political advantage—democratization has not brought the anticipated gains in transparency and accountability.[57] A small, concentrated group of elites continues to dominate the political system through political parties oriented around powerful personalities and major business interests, rather than around ideas supported by large groups of voters.[58] Although the electorate chooses who will govern, its influence over public officials and policy is limited, especially between elections.[59] The use of a closed party list system—in which voters choose a party but cannot express preference between individual politicians on the list—has further undermined the development of accountability between the political class and voters.[60]

Power has especially tended to concentrate in the presidency, which is the most developed of Ukraine's three branches of government. Through a mix of formal and informal power, Ukraine's presidents have exerted influence over the enforcement of laws (which are often arbitrarily enforced since they tend to be poorly written and/or contradictory), the administration of regulations, the media, the election process, patronage networks, and the appointment of oblast governors. The switch to a parliamentary-presidential system under Yushchenko introduced additional formal checks on presidential power, and he allowed much greater media independence and professionalism than his predecessors.[61] Such changes, however, now appear to have been temporary. Ukraine has returned to a presidential system, and Yanukovych appears to have taken several measures to consolidate his power, including, allegedly, increasing pressure on the media.[62]

At present, institutional development and power politics appear to be locked in a cycle wherein weak institutions enable individuals to use de facto power to gain influence, which they then use to perpetuate institutional arrangements that allow them to exercise their power unchecked. Thus, while institutional design is the primary mechanism for preventing the concentration of power, it is at the mercy of precisely the power politics it seeks to contain.[63] The result to date has been a government oriented around patronage, rent-seeking, and zero-sum calculations that is unable to deliver basic public goods. For their part, the Ukrainian people remain distrustful of government and demoralized.[64]

Although many of these attributes have been evident since independence, frequent institutional changes and shifts in the balances of power between elites have also caused marked variability in Ukraine's political system. The last five years are a case in point, as are the dramatic changes currently underway. In addition, other "drivers of change" interact with political dynamics and catalyze change. In the coming decade, questions of identity, for example, could feature prominently in the political sphere and become explosive if not adequately managed. Economic and demographic challenges could increase pressure on the political system to meet expectations. Ukraine's foreign policy options are also linked to political dynamics: how will evolving relations with Russia and the West affect internal balances of power? Conversely, how will political dynamics affect Ukraine's future foreign policy course?

## Identity

As a "borderland," Ukraine has wrestled with questions of identity much longer than it has been a state. Independence brought additional challenges. With only limited experience in self-government, the polity has had to develop a sense of what it means to be "Ukrainian" on a social, political, and international level, a process complicated by its internal divisions—rooted in a long history of interaction with neighboring peoples and powers—and its geopolitical position between an expanding Europe and an ever more assertive Russia. Over the course of the last twenty years, Ukraine has developed a sense of nationalism, statehood, and confidence in its sovereignty. Nonetheless, debates remain.

Ukraine's social fabric is complex. In the 2001 census, 77.8 percent of the country's population self-identified as ethnically Ukrainian, 17.3 percent as ethnically Russian, and the remainder as belonging to small minority groups.[65] While the two largest ethnic groups are geographically concentrated—Ukrainians in western and central regions and Russians in eastern and southern regions[66]—the exact boundary between them is difficult to pinpoint and increasingly permeable.[67] Ukrainians and Russians have historically influenced each other culturally (though in the last two centuries Russians have undoubtedly been the more influential). As a result, linguistic divisions do not correspond with ethnic divisions; most notably, about half of Ukrainians speak Russian as a first language. Religious differences have historically reinforced ethnic divisions, but have been a less prominent factor in recent years.[68] Given this social diversity, ideas about who is "Ukrainian" are constantly evolving.

Overall, a sense of a civic, multinational—versus ethnically based—Ukrainian identity is spreading,[69] but it remains contested in the political sphere. Political party affiliations and positions on key issues divide geographically along roughly the same lines as ethnic identity,[70] such that the country is often divided into "two Ukraines": one promoting the Ukrainian language and culture and closer ties with the West and the other advocating official status for the Russian language and closer ties with Russia. However, as with ethnicity, the boundary between the political East and West is increasingly blurred. Moreover, the most vocal ideologues at either extreme represent only a minority of Ukrainians: most Ukrainians belong to the "other Ukraine"[71]—a group with no clear boundaries that is "for the most part invisible, mute, uncertain, undecided, ideologically ambivalent and ambiguous."[72] This group's uncertainty renders it vulnerable to political manipulation and "brainwashing," which politicians readily employ in competing for its loyalty.[73] However, it also has the potential to act as a moderating force.[74]

One identity-related issue that has sparked heated political debates is history: what is "Ukraine's" history? Is it the history of the Ukrainian people? The territory? The state?[75] As Zbigniew Brzezinski has noted, Ukraine is in the process of "recovering its historical memory."[76] While vital to the development of both nation and state, this process has been controversial and polarizing, particularly as government officials—including, and especially, the president—have intervened. Viktor Yushchenko, for example, issued official commemorations on controversies surrounding Ukrainian nationalist movement leaders Roman Shukhevych and Stephan Bandera, Ivan Mazepa, and the Great Famine of 1932–1933 (known as the Holodomor), reinforcing a particular emerging national narrative.[77] Viktor Yanukovych has since criticized Ukrainian nationalism and downplayed Stalinist crimes in an apparent attempt to redefine the national narrative.[78]

Thus, although Ukraine's identity evolves slowly, it can play a dramatic role in change under certain circumstances because it is intertwined with ideas about the country's political, economic, and international choices.

Internal divisions have an upside: they make it difficult for any one political force to monopolize power.[79] However, they also contribute to political instability. Unresolved "wedge issues"—such as the official recognition of the Russian language, the status of Crimea and the Black Sea Fleet, and Ukrainian history—could escalate politically and deepen polarization.[80] Determining Ukraine's geopolitical orientation—at its core an identity question—is a particularly provocative issue, which, if managed poorly or manipulated by external actors, could threaten the country's national coherence and territorial integrity.

## The Economy

Prior to independence, Ukraine was heavily integrated into the command economy of the Soviet Union, producing one-fourth of its agricultural output and playing a central role in its energy and industrial sectors. Ukraine has since transitioned to a market economy, albeit one that noticeably underperforms.[81] The pace of market-oriented reforms has been uneven since independence, reflecting, in part, the difficulty of undertaking such reforms in an institutionally weak, politically divided state. There have been two waves of reforms, one beginning in late 1994 and another in 2000—both following crisis and leadership change. However, beyond joining the WTO in 2008, recent reform efforts have been disappointing. Privatization, for example, a critical reform for post-Soviet countries, has stopped since 2005.[82]

The stagnation of the reform process has undermined Ukraine's economic performance. The economy's reaction to the global financial crisis was one of the most severe in the world—a 15 percent drop in real GDP growth in 2009. The crisis was particularly severe in Ukraine because of the poor state of its public finances at the time (its current account deficit reached 7 percent of GDP in 2008), the collapse of global prices for steel (which accounted for 42 percent of total exports in the first half of 2008), and its exclusion from global financial markets because of its political gridlock and poor business climate.[83] Real GDP is forecast to grow around 4.7 percent this year, but this recovery has primarily been sustained by rising steel prices (Ukraine's primary export). A more sustainable recovery has been hampered by political gridlock, which has prevented the implementation of necessary reforms, including those required for the disbursement of IMF assistance funds.

The economy suffers from a number of chronic problems, such as low productivity, an ineffective regulatory system (Ukraine ranks 142nd out of 183 countries in the World Bank's Doing Business Index), pervasive corruption (Ukraine ranks 146th out of 180 countries in Transparency International's 2009 Corruption Perception Index), obsolete physical infrastructure, a large shadow economy, underdeveloped capital markets, ineffective social policies, and persistent budget deficits (in large part due to direct support for inefficient monopolies).[84] Ukraine's economy is the most energy-intensive in the world and relies on imports to meet the bulk of its energy needs; consequently, it is particularly vulnerable to fluctuations in energy prices, the value of its currency, and the interests of its main supplier, Russia.[85]

Most critically, after nearly twenty years, Ukraine's market economy is still struggling to fulfill its primary purpose: to raise Ukrainians' livings standards. Even after relatively strong growth between 2000 and 2008—an average of 7.2 percent per year—Ukraine's per capita income (at PPP, current international dollar) is $6,656, about 22 and 45 percent of per capita income in the EU and Russia, respectively.[86] Unemployment was estimated at 8.8 percent in 2009, and additional Ukrainians are either unregistered or underemployed.[87] In a recent poll, 71 percent of respondents thought Yanukovych should focus on job creation (compared to 3 percent that prioritized relations with the EU and 1 percent that prioritized relations with NATO).[88]

In the coming decade, Ukraine's reform process, economic performance, and standard of living could vary substantially, with significant effects on other spheres of Ukrainian life. Much of this variability is derived from the interaction of economic forces with other drivers of change. First, the pace and direction of economic change in Ukraine is determined in the political sphere. To date, the government's constant interference and failure to undertake critical reforms has visibly prevented the economy from reaching its potential.[89] According to the Independent International Experts Commission, "Ukraine's key problem is that the state malfunctions so much that it is unable to carry out its duties towards its citizens, while hindering the citizens from solving their problems on their own."[90] It is conceivable, however, that the level of institutional capacity for carrying out reforms, managing the economy, and distributing the benefits of growth—as well as elite influence over these processes—could change.

Second, economic change in Ukraine is influenced by the country's foreign relations. In 1991, Ukraine chose to adopt the economic model of "the West," and European and American organizations and investors continue to actively advocate deepening liberal reforms. Ukraine's economic ties with the EU have deepened even when the political dimensions of the relationship have seemed to stall. Prior to the financial crisis, the stock of direct investment from the EU in Ukraine was accelerating, increasing 75 percent between 2007 and 2008.[91] Trade between the EU and Ukraine has also increased over the last decade (24 percent of Ukraine's exports were destined for the EU compared to 21 percent to Russia),[92] in part due to deeper business ties and in part due to trade agreements, including a Generalised System of Preferences (GSP) in place since 1993, Ukraine's WTO accession in May 2008, and ongoing negotiations regarding a free trade agreement. Concurrently, Russia continues to play a prominent role in Ukraine's economy and economic policymaking, primarily through its dominance over Ukraine's energy supply and its ownership interests in major industries.[93] Moving forward, Ukraine's relations with both West and East will affect its economy: a decisive change in foreign policy orientation could alter the dynamics of foreign influence over the economy altogether; conversely, economic relations could provide the impetus for deeper partnership.

## Energy

In Ukraine, energy is highly politicized, influenced by both foreign relations and domestic politics. Consequently, a mix of private and public interests exert influence over Ukrainian energy policy. A recent independent report on reform proposals for Ukraine called the energy sector "the biggest source of waste and corruption in the Ukrainian economy."[94]

Ukraine's energy sector factors into its relations with both "West" and "East." First, Ukraine is integral to EU energy security. Ukraine is the transit country for around 120 billion cbm of European gas imports from Russia, equivalent to about 20 percent of Europe's total gas consumption[95] and 73 percent of Russian gas exports to Europe.[96] Since disputes between Ukraine and Russia over energy pricing led to gas cutoffs to Europe in 2006 and 2009, the EU has grown more concerned about its reliance on Ukraine's pipelines. Although Russia has cited these incidents as evidence of Ukraine's unreliability,[97] for Europe, they held a broader lesson of the imperative of supply diversification and a common energy policy.[98] For Ukraine, the gas crises highlighted the potential for its political disagreements with Russia to affect its economy. Since gas supplies and pipelines in both Ukraine and Russia are controlled by state-owned monopolies, political relations between their governments directly impacts the energy industry.

Second, Russia's relatively assertive foreign policy in recent years has translated to the energy sector in a number of ways: attempts to increase influence over Ukraine's energy sector (for example, Russia has proposed a merger of its state-owned gas company, Gazprom, with Ukraine's gas distribution company, Naftogaz, which would effectively amount to a takeover given their relative size);[99] attempts to undermine Ukraine's strategic position (for example, by supporting the construction of the South Stream pipeline, which would bypass Ukraine); and, most importantly, using energy as a bargaining chip (for example, agreeing to lower gas prices for Ukraine in exchange for an extension of its lease on the Sevastopol military port).

Ukraine's energy sector suffers from a number of domestic challenges, too: essentially, "in the Ukrainian energy sector private profits are made at the expense of the state budget and national economy."[100] First, Naftogaz "has long been the poster child of inefficient and nontransparent state monopolies bearing high credit risks" and is in constant need of financial rescue from the state.[101] Much of Russia's leverage is derived from the Naftogaz's inability to pay for the gas it imports. Second, the energy sector is a major locus of corruption in the Ukrainian economy.[102] Third, Ukraine's Soviet-era gas transit system is critically obsolete. Modernization of the system has been recognized as a priority for a number of years, and Ukraine has attempted to garner EU support for ventures such as a tripartite gas pipeline consortium with Russia. The Yanukovych administration has continued this effort, showing intentions to work with both the EU and Russia to improve its reliability as a transit country.[103]

Fourth, and in many ways driving all of the above, the government's practice of setting—and subsidizing—artificially low gas prices for households, state-funded organizations, and municipal heating enterprises forces significant losses on Naftogaz (about $2–$3 billion per year, which it finances through debt and can only repay with government assistance), reduces incentives for improving energy efficiency (Ukraine is one of the most energy-intensive economies in the world, with an energy intensity 2.5 times that of European countries),[104] and discourages domestic energy production.[105] Normalizing prices, which would reduce waste and improve efficiency, has been staunchly resisted by leaders of heavy industries—particularly steel, one of the country's major exports—because higher energy prices would reduce competitiveness. Nonetheless, per its agreement with the IMF, Ukraine has committed to raising gas prices by 20 percent per quarter until they reach cost-recovery level.[106] The political consequences of this new policy remain to be seen.

How Ukraine develops its energy policy in the next decade will contribute to the nation's prosperity and international influence. Domestic policy to reduce energy intensity and waste would contribute to stable economic growth. Rather than subsidizing cheap gas, the country could benefit from public investments in more productive sectors. Foreign relations with both the EU and Russia will also play a role. International funding to Ukraine from bodies such as the EU and IMF hinge on energy reforms. Simultaneously, Ukraine faces increasing pressure from Russia, which could complicate its reform process and its potential as a stable partner for Europe. Navigating a path that orients the country's energy policy toward the interests of all Ukrainians will be challenging.

## Demographic Change

Since independence, Ukraine has experienced an unprecedented demographic decline. Ukraine has the highest rate of population decline in Europe: its population has decreased by 12 percent since independence (from 52 million in 1991 to 46.2 million in 2007). If current trends continue, Ukraine's population will decline to 36.2 million by 2050.[107] A number of forces drive this trend. First, emigration rates have increased since independence and account for about 20 percent of the recent population decline. Second, Ukraine's fertility rate is below replacement levels, having declined from 1.9 to 1.2 per 1,000 live births between 1982 and 2008. Since 1991, death rates have exceeded birth rates in Ukraine; in 2007, there were 10.2 births compared to 16.4 deaths per 1,000 people. Third, Ukraine's mortality rate is rapidly increasing. Adult male mortality levels are especially high: the probability of dying between the ages of fifteen to sixty for Ukrainian males is 384 per 1,000—a level comparable to that of low-income countries. Ukrainians are not only dying younger but spending less of their life in full health compared to other eastern Europeans.

These trends point to a health crisis in Ukraine. Ukraine's high mortality rate has resulted primarily from poor management of noncommunicable diseases and chronic conditions. Mortality is linked to unhealthy lifestyles and preventable factors, especially tobacco smoking (more than 10,000 deaths per year), road accidents (more than 100,000 deaths per year), and alcohol consumption (more than 40,000 deaths per year).[108]

Ukraine is struggling to manage its demographic decline and the underlying health crisis, but social sector reforms have stalled since 2004, leaving in place a system with serious shortcomings, including disproportionately high expenditures, insufficient services, and severe inequalities in the delivery of services.[109] Demographic change and a continuing health crisis will increase pressure on the political system and heighten the need for economic stability and growth to finance reforms. In the absence of reform, the consequences of an aging, unhealthy population could become more visible. In addition, increasing regional disparities in population growth and health conditions could exacerbate tensions.

## Foreign Policy Orientation

Ukraine gained independence in the context of profound geopolitical changes. For Ukraine, the collapse of the Soviet Union signified not only shifting political boundaries, but also the need to replace the discredited ideology that had informed its institutions for seventy years.[110] At the time, "Europe stood as a beacon," presenting attractive models for achieving political stability, economic prosperity, and secure sovereignty.[111] Ukraine thus made the "European choice" of democracy, a market economy, and integration into the Euro-Atlantic community. In the course of the two decades that followed, however, this choice has become contested. It remains uncertain how Ukraine will manage and prioritize its relationships with East and West in the future.

Much of the uncertainty regarding Ukraine's geopolitical orientation has resulted from factors internal to Ukraine. First, Ukrainians are divided (often regionally) on the issue.[112] A December 2008 poll by the Razumkov Centre found that 27.5 percent of the population favored prioritizing relations with the EU in foreign policy, while 51.1 prioritized relations with Russia.[113] Second, Ukraine's four presidents to date have differed significantly in their approach to foreign relations, despite purportedly sharing the goal of balancing closer integration with the West with stable relations with Russia. Frequent changes in Ukraine's foreign policy direction—from Leonid Kravchuk's efforts to use the West as a counterweight to Russia, to Leonid Kuchma's "multivectorism" to Viktor Yushchenko's avid prioritization of integration with the West to Viktor Yanukovych's rapprochement with Russia (while maintaining that his ultimate goal is good relations with both East and West)—have caused considerable strategic confusion.[114] Third, elite infighting has prevented long-term commitment to a single, coherent goal.[115] Political gridlock has prevented Ukraine from following through on its commitments to Western institutional partners such as NATO, the EU, and the IMF,[116] while simultaneously creating opportunities for a more assertive Russia to regain influence over the country's political and economic affairs.[117]

External factors have undoubtedly affected Ukraine's foreign policy choices as well. Since the mid-1990s, the United States has maintained a robust bilateral relationship with Ukraine and argued that Ukraine as "a stable, independent, democratic country with a strong market economy and increasingly close links to Europe and the Euro-Atlantic community" would act as a stabilizing force and model of democratization in the region.[118] The United States has been instrumental in keeping an "open door" to Ukraine with respect to NATO membership.[119] Since 2003, NATO and Ukraine have created NATO-Ukraine plans of action annually, and after Yushchenko's election in 2004, talks about prospective membership gained momentum. However, Ukraine's political gridlock and weak public support for membership (only about 20 percent of Ukrainians support NATO membership, compared to 30 percent in the 1990s) have prevented the creation of a membership action plan (MAP).[120] Since his election, President Yanukovych has removed NATO membership from Ukraine's agenda entirely, dissolving six specialized structures for coordinating integration in April 2010.[121]

Europe has been less certain in its approach to Ukraine than the United States.[122] A Partnership and Co-operation Agreement between Ukraine and the EU came into force in 1998. However, subsequent agreements, such as the inclusion of Ukraine in the European Neighborhood Policy in 2004, have aimed at facilitating integration, but not the EU membership anticipated by many Ukrainians.[123] Negotiations are currently underway on an EU-Ukraine Association Agreement, which would include a Deep and Comprehensive Free Trade Area, but again, not steps toward membership. The EU's reluctance toward membership for Ukraine stems from resistance from certain member states (especially Germany and France), which are grounded in a number of concerns: Ukraine has proven itself extremely slow and ineffective in undertaking the kinds of reforms required for accession; if admitted, Ukraine would require a large portion of the EU's internal assistance funding; Ukraine's large agricultural sector could jeopardize the current common agricultural policy; and, the integration of twelve new members in 2004–2006 proved more difficult than anticipated.[124]

In large part, Russia has acted as the "pace- and frame-setter"[125] for Western engagement in Ukraine. Especially in recent years, it has vocally criticized attempts by Western governments, intergovernmental institutions, and civil society organizations to push Ukraine toward democratization and membership in Euro-Atlantic institutions.[126] Russia considers Ukraine part of its "sphere of influence," which it has actively defended, especially under the leadership of Vladimir Putin and Dmitry Medvedev. It has directly intervened in Ukrainian politics on numerous occasions (most notably, in the 2002 and 2004 elections), expressed official opinions about Ukrainian political issues (such as the status of the Russian language), and actively courted Ukrainian politicians.[127] Russia has also sought to deepen its influence over Ukraine's economy[128] and include Ukraine in the customs union it is establishing with Belarus and Kazakhstan.[129] For Ukraine, maintaining good relations with Russia while defending its own sovereignty and autonomy has proved challenging. Russian-Ukrainian relations reached a low point under the leadership of avidly pro-Western Yushchenko. Current president Yanukovych has moved quickly to restore relations, but his actions—especially the Kharkiv Accords of April 2010—have raised concerns among observers and many Ukrainians that Ukraine's independence from Russia may be at risk.[130]

It is common to assume that Ukraine "can do nothing else but fluctuate between Russia and the West."[131] Will this be the case in the coming decade? Ukraine's choices with respect to its institutional membership and foreign policy objectives will clearly drive change, as will its ability to manage the relationships it has not chosen to prioritize. In addition, evolving external conditions could alter the options available to Ukraine. What would a "Western orientation" entail if the EU remains unwilling or unable to offer membership to Ukraine? How would Ukraine's maneuvering room be affected if relations between Russia and the West were to deteriorate?

*Conclusion*

Ukraine's recent history has demonstrated that the country's political system, economy, society, and foreign relations are dynamic and often subject to unforeseen change. The drivers of change described above could vary significantly in the coming decade, suggesting that Ukraine's future course is by no means certain. Based on these variable trends and their potential interactions, what scenarios are plausible for Ukraine in 2020? Which of these scenarios do policymakers tend to overlook?

The driver research and paper were done, in the case of our pivotal country work, by graduate students at CGA/NYU, under my supervision. A variant, which I've used in scenario work for the National Intelligence Council, is to assign individual driver papers to outside experts, working within the common template described earlier. Each expert sends the paper to participants at least two weeks prior to the workshop, then presents the results during the first day of the two-day session. After each driver presentation and discussion, the entire group looks across the drivers for interesting interactions, and nominates alternate scenarios for in-depth consideration.

Trends in important drivers, and their interaction, provide architecture for the interactive discussion about alternate scenarios, a structure for the scenario narratives, a basis for comparisons across scenarios, and a means of targeting research and intelligence collection and of tracking change. But history proceeds in fits and starts, not as the smooth progression of drivers and well-conceived policies. And the future comes at us not as trends, but in the form of specific events, often surprising and consequential—what Peter Schwartz calls strategic surprises, "those patterns of events that . . . would make a big difference to the future, force decisionmakers to challenge their own assumptions of how the world works, and require hard choices today."[132] They are consequential, not in some objective sense, but because they undermine assumptions essential to current strategy or policy, and contradict our way of thinking about the world. They are surprising, in part, because the forces and antecedent events building up toward the event are unobserved or dismissed as unimportant. They are not singular discontinuities, merely perceived as such. Nor are they historically unprecedented, but may have occurred in historical contexts deemed not relevant to present circumstances. Others are surprising because they emanate from the intersection of drivers that stovepiped organizations—in academia, government, business—are not structured to notice. When future wild cards are presented to experts and observers, they are viewed as "low probability" but, when debated in the context of other, reinforcing events, gain in plausibility. Scenarios, to be credible and useful, should incorporate both drivers and wild card events (recent interest in such consequential surprises has produced multiple brands: wild cards, black swans, game changers—each with allegedly unique features. I'll use "wild cards").

Ideas for wild cards emerge from the research on drivers, and comprise part of the read-ahead material, along with drivers and potential scenarios, sent to workshop participants. Some of these ideas are already in the public discourse, and by definition are familiar, though they would still be considered by many as surprising (terrorist use of the Internet to attack targets); some occur to researchers as we consider driver interactions (will an established but increasingly fragile government survive a sudden economic downturn, a natural or environmental disaster? how might political events in one country influence events in another? how might new technology enable a sudden surge in capability and ambition of a nonstate adversary?). Others will be nominated by participants at the workshop, during the initial brainstorm about potential scenarios, or as the group works through the specific scenarios chosen for fuller development. The final scenario narratives (see below) thus combine drivers and discrete events. Both shape the story, but the wild cards lend authenticity.

At this point, roughly fifteen workshop participants have been chosen. They are experts, observers and policymakers, Americans and non-Americans, well established, and up-and-coming. The time between their selection and the workshop can then be used to prepare them for a process in structured speculation that is outside their normal experience and can produce wheel spinning and wasted time unless adjustment problems are anticipated. One source of preparation is the package of read-ahead material, consisting of a description of workshop process, the agenda, drivers paper, the list of potential scenarios and wild cards, and list of other participants. These should be sent two weeks or so before the meeting. The other preparatory step, enabled by the limited number of participants, is a one-on-one conversation with each, during which the project director describes the underlying purposes of the meeting, the unscripted nature of the conversation, the scenario construction process, and what manner of contribution is expected from that individual.

Reproduced below are the process description and candidate scenarios sent to Syria workshop participants in advance of the event.[133] As explained earlier, we did not structure the Syria meeting around drivers, given the state of civil war, potential outcomes of which trumped longer-term factors in shaping the countries' future. In most cases, the read-aheads would also include the drivers paper.

## THE SCENARIOS PROCESS: SYRIA

The process is designed as a free-flowing discussion on plausible futures for Syria over the next five years. It is not a formal simulation with assigned roles and a scripted dialogue, but a conversation about potential outcomes of the civil war. To jump-start the process on Thursday afternoon, we will begin with a quick presentation by the facilitator of some fragmentary scenarios emerging from commentary on Syria and the region (see enclosed document on potential scenarios). These tend to revolve around the political and military capabilities of the internal antagonists, the role of regional state and nonstate actors in facilitating, fueling, or containing the conflict, the political, human, and economic costs already incurred, and the shadow all of this casts over Syria's future, and policy options available to global powers and institutions.

We will then open up the scenario discussion to participants. You will be asked to react to these scenario fragments and to nominate your own. This should produce a list of several concepts. These will then be reduced to three, based on the following criteria: does the concept represent a plausible outcome by the year 2018; does the scenario depict conditions that are significant for US interests; is the scenario underappreciated in policy and academic circles, thus deserving of greater visibility; do the three concepts chosen for detailed development cover a wide enough range of variation? Likelihood—as distinct from plausibility—will not be a criterion of choice.

The three scenarios chosen will then be the subjects of detailed discussion on Friday. Here we will need to suspend disbelief and commit to building the most persuasive case we can for each scenario. To facilitate this, we will organize the discussion around drivers of change: what events and trends already evident in the present precipitate and drive the scenario; how do regional actors reinforce the scenario, and how are they impacted by it; what plausible wild card events might add momentum to the direction of change? As each scenario takes shape, we will ask about implications of the scenario for the United States (deferring discussion of US policy for later).

Toward the end of the day, we will open up the discussion to other faculty and observers for questions and comments.

## SYRIA 2018: POTENTIAL SCENARIOS

These scenario ideas are designed to jump-start our conversation on Thursday afternoon. They are derived from explicit or implied futures appearing in the commentary about Syrian civil war outcomes. They are not mutually exclusive, but each does represent a dominant tendency with distinctive results for Syria and the region. You will be asked, based on a presentation of these concepts, to eliminate the implausible or inconsequential; combine those that are redundant; nominate others that are missing from the list; and by the end of Thursday afternoon to select three that are plausible, distinctive, and that threaten US interests and challenge US policy. These concepts become the focus of discussion on Friday.

*Negotiated Settlement*: Some sort of power sharing arrangement between the opposition and post-Assad Baathist regime, overseen/managed/enforced by outside powers/institutions. It could be the outcome of military stalemate or outside pressure (regional/global consensus on greater risk than reward in continued, escalating violence/radicalization). It could evolve toward formal partition.

*Decisive Military Victory for the Regime*: With or without Assad, opposition fails to come together or fragments, increased radicalization reduces outside support for FSA while regime's supporters continue supply, and the civil war evolves into low-level insurgency but no threat to regime.

*Decisive Military Victory for Opposition*: Under outside pressure, opposition comes together politically and militarily, marginalizes radical Islamists, and enjoys enlarged support from outside actors. Regime defense progressively weakens, and Baathists abdicate. The outcome is ratified by formal settlement, with international support for reconstruction and accountability. The opposition could become the basis for effective governance (with continued, extensive outside support) or could fragment, producing another phase of the Syrian civil war.

*Regime Collapses*: Unlike previous scenario, the regime fails before the opposition coheres, producing chaos and protracted civil war among multiple actors, both Syrian and non-Syrian. The conflict continues to be fueled by arms support from regional patrons, with no faction able to gain an upper hand, leading to a de facto partition of the country along sectarian and ethnic lines, or to escalation and spillover (see below).

*Contained Civil War*: Protracted civil war occurs without spillover into region, reflecting agreement (implicit? formal?) among otherwise contending regional and global actors to avoid worst-case (below) outcomes. This "contained mess" produces military stalemate without significant escalation, possibly a slow winding down of civil war.

*Escalation and Spillover*: A stalemate develops at higher levels of violence, with expanded outside support for each side, via proxies and in some cases direct conflict among regional actors (triggered by events—the shooting down of a Turkish aircraft, a preventive strike against Iran's nuclear installations), spillover of sectarian conflict into the region, and increased great power engagement.

# The Scenario-Building Workshop

## The Agenda

Reproduced below is the meeting agenda for the Syria alternate scenarios workshop, included in the read-ahead package and the basis for introductory remarks at the meeting.

---

**SCENARIOS INITIATIVE**
**Syria Event Agenda**
**February 7–8, 2013**
NYU Center for Global Affairs
15 Barclay Street, Room 430

---

**THURSDAY**

**4:00 p.m.**  **Project Introduction**
- Center for Global Affairs
- Scenarios Initiative (funded by the Carnegie Corporation)
- The scenario process

**4:30 p.m.**  **Discussion of Alternate Scenario Concepts**
- Presentation of potential scenarios as a point of departure
- Panel review of the list of scenarios (i.e., discard, refine, and/or add new scenarios) based on the following criteria:
  - Plausibility versus likelihood
  - Significance: for Syria, the region, for the United States
  - Value to US policymakers: which scenarios deserve more attention?

**7:00 p.m.**  **Dinner**
- Optional dinner for panelists and members of the CGA Scenarios Initiative

**FRIDAY**

| | |
|---|---|
| 8:30 a.m. | **Registration** |
| 9:00 a.m. | **Participant Introductions** |
| 9:15 a.m. | **Overview of Scenarios Process** |
| 9:30 a.m. | **Final Selection of Three Alternate Scenario Concepts** |
| 10:30 a.m. | **Development of Narrative for Scenario 1** |

- Current conditions that give rise to the scenario
- Behavior of key regional actors
- Wild card events
- Implications for U.S. interests

| | |
|---|---|
| 12:00 p.m. | **Lunch** |
| 12:30 p.m. | **Development of Narrative for Scenario 2** |

- Current conditions that give rise to the scenario
- Behavior of key actors
- Wild card events
- Implications for U.S. interests

| | |
|---|---|
| 2:00 p.m. | **Development of Narrative for Scenario 3** |

- Current conditions that give rise to the scenario
- Behavior of key regional actors
- Wild card events
- Implications for U.S. interests

| | |
|---|---|
| 3:30 p.m. | **Discussion of Implications of Scenarios for US Foreign Policy** |
| 4:30 p.m. | **Q&A with Session Observers** |
| 5:00 p.m. | **Close** |

## Introduction

The introduction on the first afternoon affords the facilitator another opportunity (in addition to the process document read-ahead and the one-on-one conversations with participants) to explain the underlying purpose of the exercise, walk through the process, respond to queries, dispel misconceptions (most of which are predictable—it's not a forecasting exercise, not a simulation, not a consensus process, there's no a priori "most likely" future—see chapter 5 for "traps") and establish the facilitator's intent to actively participate in and direct the meeting (see below for role of facilitator). The introduction is essential to bring the participants from their everyday preoccupations and dug-in positions into a speculative frame of mind, prepared to suspend disbelief, entertain alternate points of view and consider futures that appear, based on current events or "history," to be implausible. Deriving value, for the participants themselves, and for consumers, depends on this mind-shift. It will begin for some participants at this point (others will be ready to go), though reinforcement will be needed as the discussion proceeds and the facilitator senses a lapse back into forecasting or set-piece positions. In making the case for the process, it's helpful to take an informal poll of the experts around the table, of who has not been surprised by important recent events in his or her country/region of expertise. Few if any will raise their hands. The contingent and surprising character of historical events, and ongoing controversies about dominant present trends, should persuade most at the table of the wide range of future possibilities and the potential value of thinking in alternate futures terms, and commit them to the purpose of the exercise. The few outliers can be brought along as the discussion proceeds, or in the worst case, will go quiet.

## Identification of Potential Futures

The process then proceeds to a discussion of potential futures arising from the interaction of important drivers (population, resources, environment, governance interactions; growth, poverty, internal and cross-border conflict; etc.), from present uncertainties, and from the literature. The starting point for this discussion is the read-ahead on potential scenarios, which is presented by the facilitator, but should extend beyond this list as participants nominate and debate other ideas. Here is the first opportunity to tap into the considerable expertise—on countries, regions, issues, US policy—assembled at the table, which will generate new scenario ideas arising from ongoing research on drivers of change, recent but underpublicized events, extant policy-driven assumptions in need of scrubbing, and theory. At this point in the process, some scenario builders suggest assembling a matrix that captures all possible combinations of drivers, then leading the group through a systematic examination of the cells for clusters of drivers and potential scenarios. In my experience, this is tedious and wasteful of limited time, and not attuned to utility for policy/strategy makers. I believe that the right combination of experts and policymakers at the table, in a free-flowing conversation about potential futures, gets to a higher level of futures insight more quickly, while still covering the important drivers, than plodding through a six-by-six matrix.

For broad questions of global change (future of international cooperation, great power relations, terrorism, global economy, access to resources, etc.), there is real value in top-down, theory-driven global scenario exercises. Our views on such macro-issues are often driven by sometimes unspoken assumptions about how the world works: liberals cannot imagine how other states can see the world and its "imperatives" very differently than we do (Secretary Kerry's palpable surprise at Putin's "19th century behavior") or sustain illiberal policies at odds with the "deep structure" of liberal globalization;[134] realists have too much confidence in local balances of power as dependable sources of US security absent US presence (imagine South Korea and Japan cooperating to contain Chinese power). If our implicit macro-assumptions about how the world works are at variance with emerging reality, our policies will be at odds with historical forces, will represent an overinvestment in a willfully self-generated narrative, and will produce diminishing returns from a finite stock of political capital. Alternate metanarratives help to identify the metrics of these diminishing returns. They enable us to do an honest risk/reward calculation on potential grand strategies.

A nice example of contrasting IR theories as source of ideas for alternate scenarios on United States–China relations is Aaron Friedberg's piece, published in 2005 in *International Security*.[135] His matrix summarizing these theory-driven scenarios is reproduced below.[136]

## Optimists, Pessimists, and the Future of US-China Relations

| *Theorists* | *Optimists* | *Pessimists* |
|---|---|---|
| Liberals | Interdependence | PRC regime: |
| | Institutions | Authoritarian/insecure |
| | Democratization | The perils of transition |
| | | US regime: |
| | | A crusading democracy |
| | | Interactive effects |
| Realists | PRC power: limited | PRC power: rising |
| | PRC aims: constrained | PRC aims: expanding |
| | Security dilemma: muted | Security dilemma: intense |
| Constructivists | Identities, strategic cultures, norms: flexible and "softening" via institutional contact | Rigid and "hardening" via shocks and crises |

These scenario fragments, numbering now between five and ten, are then fleshed out in conversation toward the end of the first day, and into dinner. Definitional issues are addressed (what do we mean by regime victory? can we imagine a peaceful outcome in Syria without power sharing; what plausible form of great power conflict could occur over Syria; what do we mean by spillover; etc.). Potential wild cards are nominated that make some scenario ideas more plausible than they first appeared. Distinctions between scenario ideas are defined. Each scenario idea is then summarized in roughly a paragraph, each in terms of the state of that country at the endpoint of the scenario, not (yet) in terms of causation or flow of events. The summaries are distributed and participants asked to consider each, as the basis for choosing three on the following morning.

## Selection of Three Scenario Ideas for Full Development

On the morning of day 2, the group is asked for reflections on the several scenario fragments, and whether any additional ideas should be considered. The collection of fragments are then assessed for significance (does the scenario engage important US interests? would policymakers see value in the scenario? should they see value, and do we think we can make such a case?); distinctiveness (is the concept redundant with any other? do the concepts as a whole incorporate a wide range of conditions?); and plausibility.

The plausibility criterion is an immediate potential source of trouble. In the many scenario-building sessions I've run, participants consistently conflate plausibility with probability or likelihood. If allowed to operate with this misunderstanding, the entire exercise can be permanently thrown off course, becoming a consensus effort to define a "most likely" future, instead of a mind-opening exercise in intelligent, policy-relevant speculation. Plausibility must be understood as imaginable, conceivable, an outcome for which we think we can make a convincing case, if we assume that certain intervening events occur. As such, this initial judgment about which scenario ideas are sufficiently plausible to be worth building out is a tentative one, subject to revision as the group begins to flesh out the scenario narrative. The goal is to arrive at three scenario concepts that are judged sufficiently important, believable, and different to warrant further examination. We are asking for trust on the part of participants, that if they choose an appropriately broad set of futures, their suspension of disbelief will be rewarded by a singular conversation and insights into previously overlooked sources of change and outcomes. In postevent reflections on the process that we elicit from participants, the debate on which scenario ideas deserve closer examination is often judged as the most rewarding of the event.

## Building the Scenario Narratives

Beginning midmorning of day 2, the group proceeds to discuss each of the three scenarios, with the objective of constructing a rough narrative for each. We do not expect a coherent story to emerge immediately from this roughly two-hour (per scenario) conversation, but to conduct a rich and multifaceted discussion on drivers, events, policy reactions, and implications within the context of a particular scenario idea, which produces sufficient raw material for the scenario staff to subsequently shape into effective narratives. This scenario construction phase requires a shift in mindset from choosing three scenarios to consider, to building the best possible case for each. In the former case, participants must think broadly about plausibility while erring on the side of distinctiveness, incorporating potential futures that demand imagination and stretch our understanding of what's possible. The latter phase requires that each scenario idea be accepted as equally valid (unless proven otherwise during the discussion). Although rough estimates of relative probability are a legitimate task *after* the scenarios have been constructed, going into scenario construction with the deliberate goal of creating a "most likely" future defeats the heuristic purpose of the exercise, and produces overconfidence in a single future that, perversely, increases the odds of surprise. The thought process embodied in the recommended approach is Bayesian; the value proposition is that scenario outcomes deemed highly unlikely from the vantage point of the present gain in perceived plausibility as participants step, event by event, into the future: thus the emphasis on applying a broad and tentative definition of plausibility as a criterion in choosing scenario ideas for detailed treatment.

Each of the three scenario discussions begins with an effort to locate the posited scenario outcome in contemporary circumstances. How might the existing seeds of the scenario develop in a direction that orients the future toward that outcome? What short-term future events are on the political or diplomatic calendar (elections; bilateral diplomatic negotiations; debt rescheduling; etc.) the outcome of which we can tilt toward the scenario endpoint? These present and near-term developments serve to launch the narrative into the future.

Drivers will form the backbone of each scenario. While the six or so drivers are common to all scenarios, their behavior singly and in interaction varies across the scenarios, and—along with wild cards—deflects the stories onto alternate paths. What are the key drivers of this particular scenario, internal to the country, regional or global? How does each shape the narrative? How do they interact to produce trend-altering events? How do policymakers/actors in this and other countries precipitate or respond to these developments? What wild cards, nominated earlier or emerging from the ongoing discussion, contribute to the narrative? What is it about this chain of trends and events that keeps the scenario on its course; in downside scenarios, what enabled deterioration (in the Syria regional spillover scenario, for example, why were great powers unable/unwilling to see and act on the preferable containment outcome?)? In upside scenarios, why did certain dogs not bark (what kept the spoilers at bay in the Syria peaceful settlement scenario?)? These questions, posed by the facilitator, are essential to fully explaining the scenario, for plausibility (to prevent the scenarios from becoming caricatures, to maintain complexity) and utility for policymakers (because both down and upside futures face impediments, are contingent, and require policy interventions of encouragement, prevention, or mitigation). They require that the facilitator think ahead of the group, understand the structure of each future, imagine how each could fail to materialize, and seek explanations for why and how the posited endpoint was reached.

As the full narrative takes shape, we can identify benchmarks for each. What particular trend developments or specific events are markers of the scenarios' development, indicators that a new world is emerging, rather than a slight variation of an extrapolated present? What are the metrics of this alternate future? What should we watch for? What questions should we be asking analysts and intelligence collectors?

This process of building out the scenario through discussion of drivers and events demands full and effective participation by experts at the table. They were chosen for the way their skills match the drivers of change, and for their reputations for imagination and willingness to engage actively in interdisciplinary give and take. Some will jump at the opportunity, while others may need coaxing. In either case, the quality of the discussion and of the final product demand that each fully participates.

## Estimating Probabilities

Thus far the process has not touched on relative probability of the alternate futures. The emphasis has been on constructing three plausible, distinctive, and policy-consequential scenarios, and keeping them in play as the panel builds the narrative for each. This process maximizes learning, comprehensiveness, and, ultimately, value as a prudential heads-up for policymakers too busy or devoted to current policy to see big changes coming. Structuring the process around likelihood or probability exaggerates our ability to accurately define the range of future possibilities and provides a false sense of confidence that we've gotten our arms around the future. It may satisfy the clients' demand for answers, at the cost of future surprises from outside the restrictive scope of probability as currently defined. Deliberately constructing a single "most likely" scenario, and then a few outliers, poses the same risks. The most likely is mistaken for a forecast, and the outliers are dismissed. The distinctive purpose of alternate scenarios is lost, while the client gets to check the box that due diligence has been accomplished.

Yet as we address the policy implications of the scenarios, and the actions appropriate for each, inevitably issues of relative probability must be addressed. No investment decision across alternate futures is defensible without understanding whether we're swimming upstream or down. Again in reference to Syria, the "best case" among our three was some sort of peaceful resolution, more plausibly through imposition by outside actors of a rough partition, than through regime change and power sharing. Yet the presence of multiple spoilers, both inside and outside Syria, and the rapidly changing and complex situation on the ground, militated against this outcome, and highlighted the quixotic nature of the Geneva process that was, at the time, absorbing much US political capital. This in turn recommended "containment" as representing the optimum combination of probability and benefit, and most deserving of US investment, but would have entailed risks, associated with enhancing US leverage, that the administration was unprepared to take.

Relative probabilities then, are important for a full policy discussion. "Relative" emphasizes the implausibility of any one scenario (including a deliberately constructed "most likely") materializing exactly as imagined. The judgment here is that some of the futures are more likely than others, either because they are more discernable in the present, or because the behavior of key drivers supports that outcome more than others. But remember (and the facilitator must emphasize this): the value is preparation, insight, awareness, and action to prevent or mitigate downsides and encourage favorable outcomes. Effective downside scenarios can (should) invalidate themselves, but will do so only if their relatively lower probability (as judged, imperfectly, by the scenario group) doesn't cause them to be taken less seriously than the more "likely." Scenarios are instruments of proactive policy, not academic exercises designed for "accuracy."

The timing of this discussion about relative probabilities is critical: it should take place after the participants have had a full opportunity to work through the Bayesian process of constructing each scenario and have, hopefully, learned something about how seemingly improbable futures could come to pass. The ideal time for this is after scenario construction, but before the policy discussion, or as part of the cross-scenario discussion described above.

## Identifying Policy Implications

Because the discussion is fairly unscripted, and includes policymakers, policy insights will surface at various points during the construction of scenario narratives. Facilitators should encourage these policy-relevant interventions, rather than insisting on adherence to the agenda. But after the scenarios are constructed there should be at least an hour reserved (don't delete) for a focused discussion of the policy implications of each scenario and of policy choices across them.

Within each scenario, several questions should be addressed. Because the scenarios were conceived as plausible and consequential, without a priori judgments of their degree of desirability (or relative probability), each will embody both favorable and unfavorable conditions, and each will entail risks and opportunities. The initial policy-related question is therefore, What is the net effect of each scenario on US policy, and more broadly on US interests? Which can we deem favorable, which not, keeping in mind that none may reinforce current policy and that both good and bad will make demands on policymakers. The second question is how the United States can encourage the scenario, if deemed favorable, and discourage its development if considered unfavorable. What causal factors (drivers, events, effects of previous policy) can be influenced in desired directions? What kind of adjustments or fundamental changes to current policy would be required in both up- and downside scenarios? The third set of questions accepts the possibility, even likelihood, that US policy will be unable to tilt a particular scenario decisively toward desirable outcomes, or away from negative ones, that policy-relevant questions are confined to how to protect interests in the scenario as depicted: how to manage/mitigate downsides, how to sustain favorable conditions, what indicators to watch, what additional knowledge we need. Policy questions should be posed across the three alternate futures, with relative probabilities as well as risks/opportunities weighed. In the case of Syria, for example, the threats to US interests posed by the plausible downside (regional sectarian war, now the reality) suggest a greater willingness to bear the costs of investing in the most plausible upside (containment), even if these costs include the use of force to enhance leverage over other actors, with its attendant risks. Finally, in some cases, despite the distinctive character of the alternate scenarios, certain important conditions may be present in all three. Chapter 1 suggested three key assumptions about the future macro-environment—relative US decline, civilizational conflict, globalization—that afford a context for evaluating the costs and benefits of alternate US grand strategies.

## Wrap-up

The last hour or so of the day-and-a-half event can be used for assessments of the process by participants (what have we learned about how to maximize insight?) and for bringing others into the discussion. In each of the sessions sponsored by Carnegie, fifty or so observers attended and were asked at this point to join the conversation. They included representatives of the sponsoring organization, faculty and graduate students not participating directly in the project, and other experts with an interest in the country or in the scenario process, including the press.

## How to Facilitate

To obtain a successful result, defined as an enriching experience for participants and a useful outcome for policymakers, expert facilitation is essential. This requires a command of the process and the self-confidence to enforce it, even as others at the table propose alternate approaches that may have worked for them in the past. It benefits from some personal connection to the participants, gained during the one-on-one discussions prior to the event, or through previous scenario-building exercises or other professional activities. It requires knowledge of the skill sets around the table, an ongoing awareness of which drivers have not been sufficiently covered, and the ability to draw out panelists who may prefer to withdraw from the active, fast-moving discussion. It requires sufficient knowledge of the country or issue to keep the discussion on target, ask pertinent questions, keep outliers in play, sense gaps in coverage of drivers or events, and credibly challenge and push experts who are far more knowledgeable on the subject but may have narrow comfort zones and tend to miss wild cards or driver interactions. It requires an understanding of where such discussions can go wrong: a dominant personality (watch out for hedgehogs!) who needs to be contained; a tendency to converge in building the narrative and to overlook—and therefore fail to account for—impediments to the scenarios' unfolding; a premature desire to think in terms of probability and to discount low-probability, high-impact events. Facilitation also requires an understanding of the policy context, of what questions are important to policymakers, of what would constitute policy-relevant outcomes. Obviously, facilitation in such sessions is active, engaged, and informed.

## Composing the Scenario Narrative

The result we're looking for at the end of the meeting is a full, rich conversation of drivers, wild cards, scenarios, and policy implications. We do not expect full-blown narratives from such an interactive session, but sufficient material to allow the scenario staff (in my case, graduate students assigned to the project and fully involved in the preparation) to construct the event-by-event narratives. A good record of discussion is, of course, essential to this. Students working on the project take extensive, but not verbatim, notes on the initial discussion about which three scenario concepts to develop, on each scenario narrative as it comes together, and on policy implications. These notes are assessed the following day, organized by scenario, then distributed to all for a read-through before the narratives are composed.

Each narrative should follow a common template, with an introductory summary on the distinguishing characteristics of the scenario; a quick explanation of the key drivers and events of that particular scenario; the chronological scenario narrative; and policy implications (the final section can appear as a separate chapter, as was done for Syria). While designed to engage the reader in a "future history," the writer occasionally steps back from the narrative to explain particularly important deflection points when the story departs from the other two, spotlight key driver interactions and events, and describe policy effects.

Reproduced below are common instructions for scenario writers, done during the Ukraine project.

---

1. Read over the notes—and not only yours!

Begin by carefully reviewing the notes on your scenario—including those taken by other people—to refresh your memory. As you read, identify:

a. **Broad themes**—what are the broad contours of this scenario?
b. **Debates**—where did people agree or disagree about the broad constructs of the scenario?

To begin with, ignore the micro-details—wrap your mind around the architecture of the scenario.

2. Describe the country in 2020.
    How does it grow from the present?
    How does it differ from its present state?
    What are its defining features?
    Calendar of events

3. Determine the line of causality that gets you to the endpoint.
    What is the gap that needs to be bridged between now and the scenario's outcome?
    What key causal links among drivers did the panelists identify?
    Wild cards
    How can these be ordered to get to the endpoint?

You should be able to come up with a line of reasoning that looks something like this:

Yanukovych effectively manipulates the system + opposition remains weak → Yanukovych gains monopoly control of political system → enables him to get things done → he uses his power to meet elites' and the public's most basic expectations → legitimacy (+ weak opposition) → reelection → need to keep meeting expectations → new foreign policy strategy

4. Determine the architecture of the scenario.
   What are the key drivers?
   When does each become most salient?
   What are the 3–4 pivotal events that give shape to
   the story?

You should be able to come up with a broad outline that looks
something like this:

| Time period | Main trends/events |
|---|---|
| 2010–2012 | Yanukovych consolidates power |
| | Opposition parties weaken |
| | Yanukovych takes steps to reduce resistance to his rule |
| | *Events*: 2012 parliamentary elections |
| 2012–2015 | Yanukovych meets elites' and public's expectations |
| | Expectations are for stability and economic growth |
| | *Events*: 2012 Euro Cup |
| 2015 | Yanukovych's approach is tested |
| | *Events*: 2015 presidential election |
| 2016–2020 | Yanukovych maintains power and legitimacy |
| | Economy grows |
| | New multivectored foreign policy bears fruit |
| | *Events*: none |

5. Fill in the gaps.
   Review the meeting notes again, looking for
   specific details that panelists offered as ideas for
   reinforcing the drivers/trends in the scenario.

Decide in which section you are going to put each. Draw on your own research to supplement any sections that need more supporting evidence.

6. Write with a purpose: causality, not just a string of events. For each section, you should have a list of the key topics you want to cover in the order you want to write about them. Ideally, this order should help you build your case for why x causes y causes z.

→ e.g., for the 2010–2012 section above:

1. Yanukovych manipulates the political system to his advantage
2. The opposition protests, but has weakened because
   a. PoR's success is overwhelming
   b. Faces internal organizational problems
3. 2012 parliamentary elections, success but shows need to boost popularity
4. Yanukovych takes measures to reduce resistance to his rule
   a. "Corrects" relationship with Russia
   b. Reduces emphasis on identity issues; new pragmatic vision for Ukraine
5. Yanukovych uses his influence over political system to fix economy
   a. Can push through hard policies
   b. Can attribute them to the quest for a new, strong Ukraine

- **Use transitions** that make the causal links: "One reason Yanukovych was able to manipulate the political system to his advantage was that the opposition had weakened."

- **Use intro sentences and numbers** to connect the pieces of your argument. "One reason Yanukovych was able to manipulate the political system to his advantage was that the opposition had weakened, *for several reasons*. First, the PoR was making its life difficult . . . Second, it faced internal organizational problems . . ."

*Remember*:

1. Anything that is causal/pivotal should be a significant, well-known, plausible factor.
   - Important events/trends *should not appear out of nowhere*; they should be linked to what has been said before.
   - The more important an event, the more *"macro"* it should be. In other words, try not to use obscure, highly imaginative events/etc. as the crux of a trend. "Micro" things should be used as examples or reinforces of what is already evident from some bigger issue.
   - Wild cards should be used to *reinforce* trends, not initiate them, as much as possible.
2. Use details wisely and sparingly.
   - Include details that materially strengthen the case for your scenario. *Don't bombard* your reader with irrelevant facts.
3. Know your audience.
   - Your target audience is *likely to know* a good deal about the country—no need for excessive background on issues/parties/actors.
   - Your audience knows you're making this all up: *be concise*. Show them you've thought about it and have come up with good reasons why it's plausible. But don't include unnecessary fluff.

- Your audience is busy: make your writing *readable*. Avoid highly "academic" language. Make your point as a clear as possible. Simplify your sentences.
4. Avoid passive sentences wherever possible.
  - *Attribute specific actions to specific people* or groups of people in order to give more "flesh" to your story. Only use passive constructions (i.e., *It became clear that . . .*) to communicate general trends that no particular group should care about; if some should care, use an active voice (*Many voters disapproved of . . .*).

The drafts are shared among the scenario staff and the project director, revised to clarify and explain differences across scenarios, then sent to all participants for their suggested revisions, overall assessment of the value of the process, recommendations for process improvements, and clearance to use their names in the final document. Participants are invited to critique the use of drivers and wild cards, the sequencing of events, and the treatment of policy implications. Since the drafts take from three to four weeks to produce, participants are asked to update the narratives and to reflect again on the plausibility, distinctiveness, and significance of each scenario. These comments are, where possible, incorporated into the scenarios, and the final reports posted on our website and emailed to our network. On-site briefings are also conducted with governments, think tanks, academic institutions, and firms.

The feedback on the process should be taken seriously and carefully considered in a postevent self-assessment. Process can always be improved, and lessons can be learned from both successes and failures. We revised the process several times during the Carnegie project, reducing the number of participants (from twenty for China and Russia, to fourteen for Ukraine, Pakistan, and Syria), increasing the duration of the meeting from one to one and a half days, cutting off debates about process earlier, getting more feedback to participants as the meeting progressed, shortening the drafting process, and extending the time permitted for participant feedback.

A good example of the importance of process is afforded by our Turkey alternate futures exercise, easily the most contentious of the several we did with Carnegie support. During the vetting process, several participants voiced strong objections to the draft scenarios produced by the the NYU team after the workshop. Extensive revisions, based on written comments and further discussion, produced another vetting, and gained the groups' consensus. The result greatly strengthened the results, and producing three robust futures, one of which closely anticipated recent electoral challenges to AKP political dominance. That scenario is reproduced below.

---

### TURKEY SCENARIO 3: POLITICAL PLURALISM

*Introduction*

In this scenario, the Justice and Development Party (AKP) gains control over all branches of government through constitutional amendments and has a seemingly unobstructed path to implementing its agenda. While this continues for several years, the gap between public expectations and the AKP's performance grows when it begins to pursue policies easily identifiable as "Islamist" and neglects crucial socioeconomic problems (income inequality, regional disparities, corruption, Kurdish resentments) and EU negotiations. Rising opposition in the form of reorganized minority parties (the Republican People's Party [CHP] and Peace and Democracy Party [BDP]) and an invigorated civil society strengthen political competition and the constraints on AKP authority.

When no party wins a parliamentary majority in the 2015 elections, the political system faces gridlock, from which it emerges in 2017 after a split within the AKP and early parliamentary elections. In the new parliament, the AKP competes with a reinvented CHP for the support of smaller parties needed to push through its preferred policies. Both parties, in an effort to win over the electorate, attempt to distance themselves from ideologically charged policies—which proves detrimental to the former AKP government's popularity—and now identify themselves as capable of pragmatically remedying pressing socioeconomic problems. With reinvigorated EU membership negotiations providing the guiding vision for reforms and with civil society actively articulating public concerns, policy priorities become increasingly clear. In 2020, public demands and external pressures have created the conditions for political pluralism to produce constructive results: more effective governance, expanded civil liberties and human rights guarantees, and more equitably distributed economic growth.

## Drivers of This Scenario

### Political Competition

Controversial AKP policies generate increased political resistance. Simultaneously, opposition parties manage to reinvent themselves after years of attrition and begin to expand their support base. By the end of the decade, the CHP competes directly with the AKP for both public support and alliances with the minority parties whose backing they need to implement policies. Since citizens of all ethnicities and religions grow more concerned with socioeconomic problems throughout the decade, political parties' platforms begin to converge, and political competition revolves around bringing about tangible improvements.

### Civil Society

Civil society plays a central role in this scenario. The AKP-led government's attempts to suppress dissent early in the decade energize civil society organizations. While such organizations represent, as always, a diverse range of views, many find common ground in their opposition to the AKP, and as they ally with other groups, they grow in strength. Their ideas reflect the growing concerns among the populace that their well-being and rights are being neglected.

## The Kurdish Question

Among those most disappointed by the AKP's policies early in the decade are the country's Kurds, who feel the party's promises remain unfulfilled. The pro-Kurdish BDP garners Kurdish votes formerly committed to the AKP and becomes a vehicle for advocating the rights of the country's minorities in general. The political success of the BDP provides a newly effective outlet for Kurds' concerns and prevents their discontent from fueling violent movements. When the BDP eventually secures a substantial number of seats in parliament, it becomes a prize ally for both the AKP and the CHP, which each offer significant concessions to Kurdish regions. The human rights, economic, and security situations of these regions improves.

## The EU Accession Process

With the EU initially distracted by its own internal issues and with Turkey concentrating on other foreign policy objectives, Turkey's EU membership negotiations stall, causing many to give up on Turkey's membership. However, by mid-decade, the EU has stabilized internally and refocuses on Turkey's accession, conditional on resumption of reform. To revive progress in EU negotiations, many politicians firmly reassert their support for membership in an attempt to distance themselves from the increasingly unpopular AKP leadership. By the end of the decade, EU membership appears a real prospect for Turkey and a powerful driver of national reforms.

### The Economy

GDP growth remains fairly strong through the decade. However, without effective policies to improve the regulatory environment, stimulate new investment, and build workforce skills, growth fails to meet public expectations for rising living standards and employment opportunities (a particularly pressing problem, given the swelling ranks of unemployed youth). In addition, the AKP government neglects its commitment to expanding commercial relations with a broad range of countries, alienating the business community. Opposition parties consequently demand improved economic management and invigoration of the reform process. As politicians reap political rewards from meeting such demands, governance of the economy improves.

### Foreign Policy

Growing dissatisfaction with an AKP strategy viewed as ideologically motivated peaks in the latter half of the decade when the stagnation of negotiations with the EU becomes a central concern, and a central point of resistance to Erdoğan's leadership. Following the 2017 election, coalition governments lead Turkey back into EU accession negotiations, but retain AKP's multidirectional strategic orientation.

### The Path to 2020

#### 2010–2011: AKP's Influence Peaks

On September 12, 2010—the thirtieth anniversary of the country's last full-scale military coup—Turkish voters approved an

AKP-sponsored constitutional amendment package by a wide margin.[137] The AKP and Prime Minister Recep Tayyip Erdoğan interpreted this outcome as an endorsement of their leadership and claimed it as confirmation that citizens wished to leave behind the military interventionism of the past.[138]

The approved constitutional amendments were implemented over the course of the following year. Many of these, such as laws protecting the rights of women and children, personal data, and collective bargaining rights for civil servants, were relatively easily adopted and measurably improved human rights and civil liberties in the country. Other amendments, such as the restructuring of the judiciary, changes to the constitutional reform process, revisions to procedures for banning political parties, and the institution of civil liability for military generals, were viewed with alarm as an effort to lock in the AKP's political ascendency.

Erdoğan maintained that these amendments were essential for making the 1982 constitution suitable for a democracy, but he faced criticism from several fronts. The CHP and its supporters criticized the amendments as an attempt by the AKP to gain power and realize its Islamist agenda. The Kurdish BDP, which had led a surprisingly successful boycott of the referendum, argued the changes neglected Kurdish interests[139] by failing to extend guarantees of rights and liberties, such as language rights and freedom of speech, to the minority. Various civil society groups, including those that supported the amendments, such as the "Not Enough, but Yes" Platform (*Yetmez ama Evet*), argued that the AKP's reform package should have included greater guarantees of pluralism and freedom and removed racist and extreme nationalist language from the constitution.[140] Overall, the opposition suspected that the party's primary objective in reforming the constitution was extending its own influence over the levers of power and silencing critics.

As the AKP spearheaded implementation of the approved reforms, its opponents' dissatisfaction grew. Increasing the number of constitutional judges from eleven to seventeen proved especially controversial. The parliament and president heavily influenced the appointment of the six new judges who were, unsurprisingly, overtly pro-AKP. Political opposition decried the erosion of judicial independence, although observers noted that the judiciary had not been convincingly independent to begin with and that, in fact, their real concern was the threat to the court's traditional role as protector of Kemalist principles of secularism and national unity.[141]

Changes to the judicial system exacerbated disagreements on another controversial subject: the role of the military in civilian affairs. By the time of the constitutional reforms, the military had already been weakened considerably, convincing many observers that its tendency to intervene in politics was a relic of the past.[142] However, military-civilian relations—and the secularist-Islamist divisions they ostensibly embodied—remained in the public eye due to ongoing trials of retired and active military officers accused of plotting coup attempts in the "Ergenekon" and "Sledgehammer" cases.[143] These trials had divided public opinion since their inception, but they became even more divisive when judicial reforms were implemented in 2011. The 2010 amendments included provisions making military personnel liable in civilian courts in cases concerning "crimes against the security of the State, constitutional order and its functioning"[144] and in preventing civilians from being tried in military courts, except in times of war. While the AKP argued that subordination of the armed forces to civilian authorities was essential for democratization—an argument supported by the EU[145]—detractors maintained that these reforms were an ill-disguised effort by the AKP to silence its critics in the military so it could pursue its own agenda unchallenged.

By the end of 2011, it was clear that by succeeding in amending the constitution, the AKP had emasculated its most formidable institutional opponents—the historically secularist-dominated Constitutional Court and the military—and secured a dominant position in all three branches of government. Its popularity still appeared remarkably resilient, and in the 2011 national elections, it extended its unbroken record of electoral success since 2002, once again winning a solid majority in parliament. Although the elections did not drastically change the balance of power in the Grand National Assembly (GNA), they were significant in that the pro-Kurdish party, BDP, managed to secure thirty seats. AKP leaders dismissed these results as a temporary aberration resulting from Kurdish voters' dissatisfaction with the constitutional reforms and predicted that they would soon return their support to the AKP.

## 2012–2015: Disappointing Performance, Growing Opposition

As the AKP approached its tenth year as dominant player in Turkish politics, challenges to its hegemonic position were growing, in part as a consequence of its own failures to meet the expectations of the electorate, in part due to its constitutional overreaching and a resurgence in the opposition.

Having gained power in the wake of a financial crisis and then positioned itself as an economic reformer, the AKP depended heavily on strong economic performance to maintain its legitimacy and popularity. Robust GDP growth prior to the 2009 global financial crisis—averaging 6 percent between 2002 and 2008—had created ongoing expectations of rising living standards and expanding business opportunities. In addition to effective macroeconomic management, meeting these expectations would require a wide array of reforms to improve the business climate, curb corruption, reduce income inequality, upgrade the education and health systems, and boost the technology sectors that would facilitate a much-needed move "up the value chain."

Although the AKP's platform had long centered on delivering such reforms, by 2012 it was clearly falling short of what it had promised. The constitutional reforms and general elections had assumed higher priority,[146] and foreign investors expressed increasing reservations about entering the market, despite forecasts of robust growth. More seriously, although annual GDP growth averaged 5 percent, unemployment remained above 12 percent—and substantially higher for women, youth, unskilled workers, and residents of eastern regions. Voters' patience was growing thin with the government's (much-touted) job-creation programs underperforming and employment opportunities remaining insufficient to accommodate the country's burgeoning working-age population.

The negative effects of stalled economic reforms were compounded by several AKP missteps that fanned its critics' worst fears—that Erdoğan and the AKP had authoritarian and radically Islamist designs. In 2012, following through on Erdoğan's promises, the government attempted a complete overhaul of the constitution in an effort to bring it in line with prevailing models—namely, those of Europe and the United States.[147] When Erdoğan announced his intention to push for the replacement of the parliamentary system with a presidential system—a radical change, understood by many as playing into Erdoğan's personal plans to eventually become president—a firestorm of criticism erupted. Restrictions on the media were tightened in an attempt to contain the debate, but this only served to further radicalize and harden the positions of those who felt their views were being suppressed.

Resistance to the rewriting of the constitution surprised the AKP, which had expected that, as with the referendum of 2010, it would be able to override criticism and win support for its proposals. Most surprising for the leadership was that its own ranks only expressed weak support. This signaled that a shift was underway within the party. Disillusioned by the leadership's neglect of economic reforms and increasingly heavy-handed tendencies, many moderate voters and politicians (whose support for the AKP rested on its image as the country's best hope for democratization and prosperity) had begun to distance themselves from the party, either joining the opposition or simply withdrawing from political debates.

Conservative, Islamist voices were also growing stronger within the AKP. Policies began to be framed in ideological terms and included various explicitly religiously motivated measures, such as the lifting of the headscarf ban at universities and increasing the taxation of alcoholic beverages. Unfortunately for the AKP, this strategy did not necessarily guarantee broader popularity, as even pious Turks were primarily concerned with lingering unemployment and the poor quality of public services, especially health. Consequently, the government relied ever more heavily on patronage as a political tool, causing the level of corruption and cronyism to escalate further.

A priority shift was also evident in the government's foreign policies. Ahmet Davutoğlu, who remained foreign minister, continued to espouse the government's commitment to a "zero problems with neighbors" outlook. However, skeptics (who had long accused the AKP of Islamizing the country's foreign policy) could point to the concentration of the foreign ministry's efforts on deepening relations with Iran, Syria, and Iraq. When combined with the prime minister's habit of playing the "Islamic card,"[148] this trend led many to believe that the country's foreign policy was becoming distinctly "Islamic." Among the most disappointed were Turkey's business leaders, who, while welcoming deeper trade relations in Iran, Syria, and Iraq, felt the government was neglecting crucial negotiations with other regional powers, such as the EU, and corresponding projects, such as the construction of the Nabucco pipeline. A slow response by the AKP to the revolt in Libya in 2011 only compounded criticism toward its foreign policy.

As dissatisfaction with AKP policies grew, opposition parties found the means to reestablish themselves in Turkish politics. By 2013, the revival of the CHP under the new leadership Kemal Kılıçdaroğlu was well underway. Disappointed by its poor performance in the 2011 elections, the party had undergone a much-needed process of introspection and reorganization. Recognizing widespread disappointment with the AKP's economic policies, the CHP had developed an economic platform that emphasized social democracy. While few believed the CHP could become the next "reformist" party, its prioritization of socioeconomic issues and seemingly less corrupt management played to its advantage. In addition, under Kılıçdaroğlu's leadership, the party's support for EU membership strengthened.

The most significantly transformed party was the BDP, whose popularity had increased immensely. Its new role in Turkish politics stemmed from the spectacular failure of the AKP's management of the "Kurdish question." Enthusiasm for the AKP's "Kurdish opening" launched in 2009 had waned as early as the 2011 elections.[149] The Kurdistan Workers Party's (PKK) unilateral ceasefire announced in late 2010 was broken soon after the elections, and military and federal police presence in the Kurdish-majority regions of the country grew dramatically. At the same time, Kurdish demands for increased freedom of speech and assembly remained unmet and the socioeconomic consequences of the escalating violence—such as widespread displacement of Kurdish families—remained unaddressed. When the AKP's proposals for the new constitution did not remove the language of Article 166 (which stipulated that all inhabitants of Turkey were Turks), Kurdish voters became convinced that the party would neglect their concerns indefinitely. As a result, their withdrawal of support from the AKP, evident in the 2011 elections, became permanent. Some—especially the children of internally displaced families who had grown up in Ankara, Istanbul, and other bigger cities—took up arms against the Turkish state,[150] guaranteeing that the situation would remain explosive for the foreseeable future. Most, however, committed themselves to increasing their political voice through the BDP's representation in parliament. Thus the BDP saw its ranks swell while it attracted support from non-Kurdish center-left voters and other disappointed former AKP constituents.

Growing political opposition to the AKP was magnified by new activism within civil society. Individuals with strong opinions on Turkey's socioeconomic problems and the role of religion in politics were organizing into civic groups, many of which had links to political parties. While media censorship remained a well-practiced AKP tactic, cracks in the party's support gave dissenters a feeling that change was possible and that the rewards from challenging the AKP were increasingly worth the risks.

This growing opposition found common ground in their determination to address problems of swelling youth unemployment, deteriorating quality of health and education, the poor condition of low-income urban neighborhoods, the frequency of prison abuses, and skilled labor shortages. As civil society organizations developed their ideas and improved organizationally, they gained supporters. The youth and student movement "Genc Civiler" (Young Civilians),[151] for example, grew in number and strength by advocating the idea of a nonauthoritarian, pluralist society. Having achieved visibility during the constitutional reform referendum, the "Yes, but it's not enough" campaign turned its attention to broader advocacy of democratic reform, gaining public support from intellectuals like Orhan Pamuk. But by far the most active civil society organizations were those working to secure the rights and freedoms of Turkey's minorities, including Kurds, Alevi, Armenians, Christians, and Jews, who managed to gain national attention for their causes (although certainly not consensus around their views), in part through well-known (if underreported) attempts by federal authorities to shut them down.

In the lead-up to the 2015 elections, it was clear that political competition was intensifying, and that the AKP lacked the means to stem it. Election campaigns were intense and the journalists covering them ever more defiant of the government's threats to punish those who criticized the ruling party.

External change was also underway: the EU, having recovered from the internal struggles evident early in the decade, began showing renewed interest in Turkish accession. It reiterated the concerns of earlier in the decade—that the reform process had slowed compared to the 2002–2005 period and that democracy was eroding.[152] Perhaps inspired more by worries that a destabilized Turkey could endanger Europe than by an overwhelming desire to see Turkey as part of the EU, European officials urged Turkey to refocus on democratization and economic reform in return for renewed prioritization of membership negotiations.

## 2015–2017: Political Stalemate Ends in Pragmatic Compromise

The 2015 elections took place in the context of intense political competition and external pressure to tackle the country's mounting challenges. The campaign period was rife with speculation that the AKP would not secure the majority—or even the plurality—of seats.

In the end, the AKP lost a substantial number of seats, but remained as the strongest party in the parliament. The CHP and BDP both gained seats, as did independent candidates of other parties. Only the Nationalist Movement Party's (MHP) representation remained more or less unchanged. Without an outright majority for the first time since 2002, the AKP was not able to act without the support of at least one additional party. Because its relations with opposition parties had deteriorated so sharply in recent years, it was not clear with whom it could partner. With no grand coalition, Turkey fell into acrimonious political stalemate.

A multiplicity of views and agendas gained currency in the new parliament. The BDP pushed for public investment in Kurdish regions and guarantees for human and civil rights; the CHP advocated retrenchment of the AKP's "Islamist" policies and greater attention to socioeconomic problems; the MHP focused on wooing nationalists in the AKP and the CHP. The AKP was on the defensive. Parliamentary debates were long and tough, often ending without consensus. Although debates were often dominated by radical voices, the incentives for each party to appear more competent and relevant were strong enough to help moderate, pragmatic politicians to begin to gain prominence.

Political stalemate finally ended in 2017 with a split in the AKP and early elections, both of which were triggered primarily by renewed argument over the Cyprus conflict. The AKP split was precipitated by Ali Babacan, deputy prime minister responsible for the economy and chief negotiator in Turkey's EU accession talks, who openly turned against Erdoğan and accused him of stalling the EU accession process by neglecting the Cyprus question. AKP politicians who had been looking for an opportunity to distance themselves from Erdoğan seized this opportunity to break with him on this politically sensitive issue. They declared that Erdoğan and his supporters had willfully undermined Turkey's EU membership prospects by focusing on their personal ambitions instead of resolving the Cyprus issue and implementing the political and economic reforms needed for membership. Babacan also found support among many politicians in the BDP and the CHP, since both parties had made EU accession a central component of their platforms in the 2015 elections. As AKP MPs shifted their support to Babacan, Erdoğan's support base came to rest on staunchly conservative Islamist MPs.

The disagreements emerging within the AKP received significant media attention, and censorship wilted. The public had begun to take an eager interest in the dynamics of this new competitive environment, wondering how it would affect daily life. The details of parliamentary debates were widely disseminated, increasing pressure on politicians to make cogent arguments. Babacan's criticisms paved the way for a flood of discussion about the relative merits of the AKP's policies in recent years. The "Islamization" of public and foreign policy came under heavy fire, both for being a distraction from pressing socioeconomic matters and for undermining religious freedom.

Parliamentary debate on the opening of Turkish ports to Cypriot vessels grew extremely heated. When they reached an obvious impasse, Erdoğan called for a vote of confidence, which he lost, leading to early elections. Election results revealed that the AKP's support base had shrunk substantially and that its losses had benefited a wide range of parties, from small radical Islamic parties that attracted voters dissatisfied with the shift of policy debates away from Islamic policies to larger opposition parties, especially the BDP and the CHP, who seemed to offer voters more convincing approaches to tackling Turkey's problems.

### 2017–2020: Pluralism Drives Democratic Deepening

Undoubtedly the most significant outcome of the 2017 general elections was that the CHP found itself roughly on par with the AKP in terms of parliamentary seats for the first time in more than a decade. Both parties held more than 180 seats. While observers expected the deadlock of the previous parliamentary session to be repeated, they were pleasantly surprised that the relatively equal positions of AKP and CHP had positively changed the dynamics of competition in parliament.

The AKP and CHP challenged each other to win minority MP support (especially from the BDP, the largest minority party) in order to push through their preferred policies. Consequently, ideologically charged rhetoric gave way to more moderate policy-oriented discussions. The AKP's internal divisions had enabled reform-oriented politicians to assert a dominant position in the party, and, under the leadership of Babacan, it decisively returned to its former program of pro-EU policies and market liberalization. Meanwhile, a new generation of politicians had asserted itself within the CHP, solidifying the party's new image as a social democratic party that emphasized the democratic aspects of Kemalism over its polarizing ideological aspects. At the heart of this change was a realization on the part of each party that—given the severity of Turkey's socioeconomic challenges, the renewed lure of the EU, and the high expectations of the public—their relative performance in the coming years would determine their fate. The rapid shifting of supporters between parties in recent years had convinced politicians that no sector of the Turkish public was beyond their reach: issues such as employment, health, education, security, and EU membership resonated with voters from all religions and ethnicities. While parties' overlapping goals frequently caused bitter disputes among rival MPs over credit for successful policies, the net effect was to improve governance because no party could afford to be seen as opposing the publically popular reforms being undertaken. The relentless involvement of civil society in national political debates helped maintain pressure on political parties to perform.

Babacan, as the AKP's new leader, presented himself as chief advocate of Turkey's EU accession. In 2018, the first Cypriot freighter docked in a Turkish port, signaling not only that resolution of the Cyprus conflict was possible, but that Turkey was ready to reopen the frozen chapters of its EU accession negotiations. Significant obstacles remained, of course, but EU membership appeared an achievable rather than aspirational goal for Turkey.

EU pressure helped to shape politics. Anticorruption measures eventually improved the quality of public services and the business environment generally, which leveled the playing field for the country's entrepreneurs while raising foreign investment. When GDP growth reached 7 percent in 2019, it seemed the government's new reform orientation was paying off. In addition, the EU's insistence on guaranteeing human, civil, and minority rights—combined with the pivotal role of the BDP in parliament—fostered a marked improvement of the government's approach to the Kurdish question. Public investment in infrastructure and services in Kurd-dominated regions increased rapidly, creating hopes of a more prosperous future. Language rights for Kurds expanded as well, with many official forms and documents available in Kurdish. Other minorities also benefited from this new approach, particularly as a result of several measures to increase religious freedom.

While reengaging with the EU, successive Turkish governments maintained the AKP's earlier emphasis on improved relations with all its neighbors, and with powerful states outside the region. The economic and strategic opportunities for large, rapidly growing countries such as Turkey had expanded in a more multipolar world, and Turkey's competitive democratic politics and improved EU prospects had positioned Turkey to seize these opportunities. It could now credibly position itself as a gateway between East and West, between the Muslim Middle East and the secular states of Europe and North America. Attractive to the East for its economic and political access westward, and to the West as a successful, moderate Muslim democracy, Turkey could now reap benefits in both directions.

By 2020, Turkey's political landscape was dramatically different than in 2010. The polarizing tensions that defined the political system earlier in the decade—between secularism and Islamism, between elites and the masses, between the majority and minorities—had given way to a greater diversity of debates on a wide variety of issues. As incentives for cooperating with opposition parties increased, politicians found common ground in advocating pragmatic policies. A robust, diverse civil society played a crucial role of communicating voters' policy preferences to politicians. Ideological differences and radical, polarizing views did not disappear but were marginalized. As the decade drew to a close, Turkey faced a bright future.

## Implications for US Interests

This represents the most favorable scenario, both for Turkey and for the United States. It describes a moderate politics and a pragmatic/realist foreign policy, devoted to maximizing Turkey's influence in its region and beyond, but aware of its own limitations and vulnerabilities, and prepared to partner with the United States on at least an ad hoc basis to address threats and create conditions favorable to its interests. A moderate, pluralist domestic politics arise not from an implausible self-restraint of a dominant AKP, but from a competitive political process ignited by diminishing returns to AKP policies, both domestic and foreign, the revival of the MHP and other parties, and increasing civil society resistance to AKP's Islamic agenda, a competition that rewards compromise and pragmatic problem-solving.

This more competitive politics has seemingly contradictory effects on Turkish foreign policy, and on the US relationship.

Foreign policy becomes more a product of domestic politics—among parties, and between parties and civil society—and less the expression of a dominant AKP grand strategy. This implies incoherence. Yet a strengthened liberal politics positions Turkey for renewed EU accession negotiations (during the latter part of the decade), reduces friction with the United States, contributes to a successful "Kurdish opening," and permits an effective execution of "peace with all neighbors" approach, which is indeed the most rational posture for a country with Turkey's size and position. As the door to EU membership reopens, Turkey's economic opportunities grow, and its appeal to Arab states as an avenue of diplomatic and economic access to the West is enhanced. With its Muslim population, pluralist politics, growing economy, and positive relationships with most regional and global powers, it is able to fully realize its potential influence.

This is most certainly a positive outcome for the United States, as it positions Turkey to reinforce regional stability, help in managing specific conflicts in the region, provide an example and material support for more effective governance in the Muslim world, respond to common threats (rising Iran, spreading terrorism, chaos in Iraq, Afghanistan, Libya) through ad hoc cooperation or through NATO, and generate commercial opportunities for US business. These common interests become more compelling as accelerating change in the Arab world produces potentially damaging consequences for both countries.

But these opportunities could be easily squandered by unrealistic US expectations for a "liberal" Turkish foreign policy. Turkey, as it emerges toward the later part of the decade, is successful, stable, self confident, with a pivotal position in a critical and transforming region. The substance of its strategy will remain the maximization of its power regionally and globally, in circumstances that offer great opportunity. While its threat environment will sometimes create common interests with the West, and with the United States in particular, it will not operate as a surrogate. Whether it serves as a "bridge" to the Muslim world will vary with the messages we're trying to deliver. Cooperation in confronting threats will depend on the circumstances, and on both states' willingness to compromise on issues they will often view differently. One should expect potentially conflicting responses to containment/deterrence of Iran, questions of outside intervention in the evolving revolutions in the Middle East, managing potential turbulence in northern Iraq as the US presence recedes, and the substance and process of any Israeli-Palestinian settlement. These issues also complicate US-Turkey relations in other scenarios. Here Turkey is not a precipitator of conflict and insecurity and has important leverage to bring to regional stabilization, but acting in a cooperative, or at least mutually reinforcing, way will require a long-term view of our common interests, a degree of patience, and adept diplomacy on both sides.

One area of likely conflict in this scenario arises from the increasingly mercantilist character of Turkey's foreign economic policies. This is a product of the democratic nature of Turkey's foreign policy decision-making, its robust growth and its globalizing commercial and financial interests. Competition for markets, capital, and resources will frequently threaten to overwhelm cooperation with the United States and with other liberal trading states. Turkey will find common purpose with other rising powers that question the legitimacy of the liberal trading system. EU conditionality may curb the worst excesses in commercial practice, but the EU is just as likely to accommodate these practices in the interests of membership.

Getting the most out of a relationship with a strong, independent regional power with a Muslim population and competing economic interests will be difficult, often producing unsatisfying compromises and agreements to disagree. But the combination of internal pluralism, external pragmatism, and growing power makes this the most favorable scenario for the United States. Here again, our contribution to its development will be modest, but not trivial: supporting the EU accession process; working to build regional security by encouraging Turkey's mediation efforts in Middle East conflicts and in stabilizing Iraq and Afghanistan; leveraging Turkey's synthesis of Islam and pluralism in shaping political change in the Arab world; and, when necessary, by pursuing our interests vigorously and independently (Iran?), counting on multiple common interests and effective diplomacy to contain the damage.

## Some Lessons Learned

In looking back over the several scenario-building workshops on pivotal countries, the most consistent process error has been a failure to push the group far enough beyond its collective comfort zone, to entertain scenarios that would have represented fundamental and challenging departures from the present. Ukraine was chosen, in 2009, as one of four countries to study, precisely because of its internal divisions and perilous geostrategic position. Yet when it came to selecting alternate Ukraine scenarios, the group could not imagine subsequent events as they have in fact unfolded. Similarly, the Syria group, meeting during a period of regime retreat, rejected an Assad victory as implausible, a serious miss that a different group, or more assertive facilitation, might have avoided. There is a trade-off between keeping the group fully committed to the process, and producing value-added results, and the error in these instances was to lean too much toward group buy-in. This choice will continue to challenge facilitators, risking rebellion in the interests of an expanded understanding of what's possible. The more legitimate the alternate scenario process as perceived by participants and facilitator, the more room for risk-taking.

If the process is done well, with the right combination of skills, styles, responsibilities, and expert facilitation, the value begins in the conversation: the insights, the connections between trends and events, the plausibility of outcomes previously dismissed, the policy impacts of unexpected but believable black swans, and the creation of a network of converted scenario advocates.

I have described this approach to scenario construction as primarily bottom-up. Scenario ideas emerge from the behavior and interaction of drivers and events, supplemented by alternate futures appearing in current debates about the country or issue. The scenario panel assesses these ideas for plausibility, distinctiveness, and importance, selecting for in-depth treatment three that reflect both top-down and bottom-up thinking. The author has used this approach in work for the US intelligence community, which has the difficult job of producing analysis relevant to policy, without being explicit about policy contribution to or consequences of potential futures. Longer time frames call for such an approach, given expanded indeterminacy and widened policy choice. The approach may miss future developments relevant to current policy, but is more likely to capture potential structural change and policy-forcing developments.

"Top down" approaches target policy impacts and choices more directly. Here the process may begin with examination of assumptions underlying current policies that fail to explain unfolding events, and testing of those assumptions against alternate, plausible futures. It may begin with selection of potential scenarios overlooked or willfully ignored by current policy as politically incorrect or too challenging to contemplate. It may start with identification of policy-relevant questions prompted by new events, policy reviews, or imminent negotiations. Such an approach lends itself to extemporaneous exercises occasioned by new information or policy reviews, and may succeed in producing results more directly relevant to current policy.

## Traps

Given accelerating change, and our inevitable engagement in this world, the alternate scenario process must be continuous (see next chapter). The impulse to react to strategic surprise by launching a big scenario study should be obeyed, but should also be leveraged to jump-start a more sustained evaluation of fast-changing reality. The "findings" of megastudies can be insightful and persuasive but degrade rapidly unless refreshed and applied to policy choices on an ongoing basis. The longer-term purpose is to inculcate an alternate futures way of thinking that operates even when things are going your way, and bringing this to bear on immediate decisions. So the first warning is to commit to long-term process, not to feel comfort that doing an alternate futures study covers your risk.

Do not define the issue prematurely or too precisely. Organizing alternate future thinking around a tangible, current policy question makes it easier to accomplish and promote results as relevant, but narrows the scope and can sacrifice value over both the short and long term. If you must, define the key question broadly and be prepared to redefine it as alternate futures emerge, some of which will reveal your original question as irrelevant or misleading. Our present biases are very much in play in defining the issue, and could set you up for surprise by building overconfidence that you've done the due diligence, when in fact you've retained the same blinders.

One manifestation of this current agenda-driven tendency is to focus the study on alternate policy options. This is where policymakers often want to drive the discussion, toward immediate engagement with the inbox, short-circuiting the less immediately gratifying, but more valuable, process of unfettered thinking about the future, then considering policy implications. Focusing on alternate options for addressing current policy issues is exactly what not to do in a serious futures study, conveying the impression of having covered all possibilities when, in fact, the future external context for policy has been held constant.

Do not peer too far into the future. Appropriate time horizons will vary with the subject or issue, but even broad, systemic rethinks should be proximate enough to engage policymakers. Remember, it's a fallacy to believe that fundamental change only happens over the "long term"; it's already in motion and could burst upon us tomorrow.

Don't invite scenario-building participants from the same network. Do the research on drivers, and as you identify useful and informed perspectives, then begin issuing invitations. It's easy, depending on your contacts, to assemble a panel of insiders, and some are essential for credibility and subsequent impact, but too many of the usual suspects will encourage grooved thinking and current-agenda-driven results. A group of participants who are new to each other adds to their and your experience and network, encourages a learning style in the discussions, and produces insights from cross-disciplinary dialogue and new research findings. Don't invite just hedgehogs or just foxes; you need the mix. Avoid pundits, who have too much invested in overconfident forecasts. And don't invite based on full representation of interest groups or stakeholders, unless you want to replicate already well-worn national debates.

Don't role-play. It's not a simulation, in which participants take assigned roles and play out a given contingency. Such events have value, but not in imagining alternate futures. The scenario construction process is a conversation among diverse views, skills, and cognitive styles, a structured brainstorm focused on plausible futures and their consequences. Participants make their best arguments for a given future, then suspend disbelief to make the best case possible for other scenarios.

Don't impose too much structure. The process I described in the prior chapter is essential for reasons of transparency and replicability, but facilitation must allow unexpected insights to play out: serendipity is as important as checking the boxes. The conversation should appear unscripted, not a forced march through the agenda.

Don't allow the scenario space to narrow too much. Plausibility as a criterion for choosing scenarios should be construed broadly, then narrowed if appropriate, after ideas have been fully developed. The Bayesian learning process that I've described will work, but only if the "scenario space" encompasses ideas initially considered by some as very unlikely. And if the project is done well, you will find yourself out in front of at least some of the "surprises." My major regret over the four years of our alternate futures work at NYU was not pushing the Ukraine group more assertively toward the scenario that caused me to propose Ukraine as a subject, but for which the group had no appetite. On the other hand, the conclusions of the studies on Iraq (the incompatibility of democracy and stability, the virtues of a national unity dictatorship) and Syria (spillover into regional sectarian civil war and the accumulating costs to the United States) resonate.

A good scenario is not a snapshot of conditions at some future date. For purposes of learning, and for credibility of results, the conversation and the final product should be in narrative form, with events and drivers interacting chronologically as the outcome takes shape. The learning, both within the group and outside, comes from walking into a future that gains in plausibility as events occur and causal connections are made. A snapshot is inherently lacking in credibility. Nor should the scenario endpoint describe some set of steady-state conditions (either good or bad). The scenario must, of course, exhibit overall end-conditions that are distinctive from other scenarios, but will contain as much ambiguity and apparent contradiction as the present. History will continue beyond the end date of the scenario.

By the same token, beware "internal consistency." That's a source of myopia and deprives decision-makers of a sense of where the policy choices are. History is replete with "inconsistencies," paradox, contingency, and perverse effects of policies that seemed like good ideas at the time. The scenario narratives must not be conceived or constructed as smooth extrapolations from the present into the posited endpoint, imparting the impression of inevitability. They may be moving along on one track, then veer in response to wild card events. In our Syria study, for example, the negotiated settlement scenario described a successful great power effort to bring most of the parties to the table, only to energize the spoilers, with consequences more negative than a seemingly more modest strategy of containment.

Don't look for consensus. It's a process that, when choosing and building scenarios, requires some intellectual combat. The end product is alternate, distinctive, consequential stories, not a consensus future that will extrapolate the recent past and as such be both more likely than the others and of no—perhaps negative—value in dealing with uncertainty and surprise.

Don't rush the process. It takes time to acclimate the group, clarify purposes, encourage full participation, make mistakes and learn from them, and build the right style of intellectual playfulness. Beware the matrix, which provides a false sense of security but is overly mechanistic and can waste time on implausible or redundant scenarios and reduce serendipity.

Don't lapse into a forecasting vocabulary or mentality. Closely scrutinize "that can't happen"; shift to "very unlikely, but if a and b happen, I can imagine it." Don't estimate probabilities (a form of false precision). Assess relative likelihood only after each scenario has been given its due. Remember the purpose: not to predict the unpredictable, but to manage uncertainty by encouraging appropriate policy responses, some of which are designed to invalidate the scenario.

Don't create a priori positive or negative scenarios, or jump to conclusions about "upside" or "downside." Some scenario effects will be surprising. Part of the learning is precisely in coming to a new understanding of the consequences of future events, which on their surface may appear inherently good or bad. There's very little in international politics that is enduringly either, and even the worst set of circumstances can have a silver lining, not to mention competitive opportunities for those with the right responses.

# Chapter 5

# Future Applications and Policy Process

Chapter 2 of this book looked retrospectively at policy errors that might have been mitigated by an alternate scenario process, one taken seriously by policymakers. Looking now toward the future, what emerging issues lend themselves to such a process, and what policy value might we realize?

Selecting topics for alternate scenario treatment requires an understanding of current US policy, and the type of future knowledge essential to getting it right or, if necessary, changing course. It also requires a sense of emerging trends and potential wild cards that may confront policymakers with challenging and, in some cases, unprecedented events that could invalidate extant assumptions. Policy-driven scenarios have been described as "top down," in that they begin with a current or emerging policy problem, either recognized by the policy community or not, then build out to incorporate longer-term trends and events that either reinforce or undermine existing paradigms. Ideally, they are prudential, done before trouble appears. They ask, What could go wrong with our current approach? How will we know when/if we're reaching diminishing returns? How do we manage downsides inevitably associated with current strategy? Or scenarios may be initiated when current policy is shocked by some consequential, unexpected event.

"Bottom up" studies can begin with questions about countries, regions, global trends, or specific issues that are subject to rapid change and that may create an altered context for current and future policy. They begin with broad questions about drivers, events, and potential alternate futures, then work back to existing policies and mindsets at potential variance with the world as it's becoming. They score higher than top-down studies on inclusiveness of scope and are more likely to spot big changes from forces invisible to those worrying about the current agenda, but run the risk of being dismissed as unserious, deliberately iconoclastic, irrelevant. In our recent work on pivotal states, we've tried to combine these approaches but have erred in the direction of maximum scope.

## Some Suggested Studies

One could imagine several top-down, policy-driven scenario studies of value to this and the next administration. One would examine resource (broadly defined) availability as a source of restraint on future American grand strategy. The debate among foreign policy experts is all over the map on this, with assertions of decline or renewal deployed to reinforce a priori foreign policy preferences, often without the skill sets necessary for credibility. Better to assemble a mix of policy-oriented economists, scientists, engineers, and students of American politics, and develop credible, alternate projections of the underlying determinants of American power, then ask how these scenarios will/should influence grand strategy when viewed in the context of the rise of China, India, or other high-population, high-growth countries.

I suspect that such an assessment would see real promise in America's competitive future, derived from its healthy demography, economies of scale, robust private markets, abundant venture capital, legitimate political institutions, geographic advantages, and hard power. But it would also project relative decline, as emerging countries grow faster and seek to leverage their newfound capacity. This suggests future multipolarity, with the United States first among equals, unable to dominate the system or impose its will in regions of still prime importance, but uniquely able to protect its security from a multitude of threats that bedevil others less advan-taged by size, internal stability, external power, or geography, and still uniquely positioned to assemble ad hoc coalitions to deal with global threats as they arise.

This alternate, but likely future, rules out "primacy" as exces-sively demanding of resources and unlikely to generate sustain-able success (see Iraq). It also rules out liberal order building (see Doha, the new China-sponsored Asian Infrastructure Investment Bank, Ukraine), a uniquely American project with a shrinking constituency and diminishing returns. It suggests that Obama had a good idea when, having absorbed the lessons of Iraq, he articu-lated a strategy of restraint and rebalancing. Unfortunately, he overcorrected. His failure to see and react to trouble, to manage the downside risks of retrenchment, has been his downfall, not the conception itself. Let's hope that this failure doesn't produce a pen-dulum swing back to a willfully inflated view of American power and interests.

Jeffrey Legro has made a similar point: "The US embrace of AIM (American Internationalism, embodying global leadership, military superiority, support for democracy, free trade, multilater-alism) is likely to be increasingly at odds with the emerging condi-tions the country will confront in the years ahead. At a minimum US leaders will need to reorient public expectations for AIM and recognize the need to cut deals, in light of pressing problems, that may involve more give from the United States than has been neces-sary since World War II."[1]

A second scenario study idea would be designed to fill a large gap in neorealist advocacy of restraint, namely how to get from here (waning primacy) to there (secure, but restrained). It would represent an effort to save the concept from its growing list of detractors, including some from Obama's own administration, who correctly observe an inflexible commitment to restraint, regardless of the signals it sends and the future liabilities it accu-mulates. The organizing question is how to manage the process of dialing back on commitments, when the world doesn't permit the fantasy that your allies will step up, or that enemies won't try to take full advantage. How do you manage expectations? Where and how do you draw the line, without being "sucked back in"? How do you avoid free-fall? How do you fully leverage the power you still possess in relative abundance, to prevent the world from collapsing around you? The study would imagine alternate scenarios depicting future challenges to US interests (from other great powers, ISIS, climate change, global economic crisis, pandemics) and work through US responses within a multipolar order and an overall strategy of restraint.

I would also take a look at the intelligence, analytic, and decision-making requirements of sustaining a strategy of restraint. With relative decline and lowered ambition, making the right moves at the right time becomes an essential element of power. "Don't do stupid stuff" is necessary, but not sufficient; acting effectively, if not literally in advance, then in the early stages of new challenges, becomes a "force multiplier." This is true whether the challenge is threatening or offers opportunity.

I would take a longer-term look at several current regional situations. The Middle East map is being violently redrawn, and efforts, diplomatic or otherwise, to preserve it seem increasingly quixotic. Bombing ISIS to save the Kurds is both necessary and gratifying; so is replacing Maliki. But the problem is massively greater than these extemporaneous responses suggest, and investing in low-probability projects—even if the amounts are modest—is less and less defensible given limited resources and the risks of failing to address root causes. The anonymous "senior White House official" had it exactly wrong when saying to a *New York Times* reporter, in late August 2014, "The way we're thinking about it (the dissolution of Iraq) is, we've got to put out the fires first. Then we can think about the future of Iraq."[2] It's essential to get out in front of Middle East turmoil by imagining the region without its externally imposed borders, then asking how we—with others—can make this likely outcome less violent, and if this looks to be impossible, how to protect our interests as the region descends into chaos. If it's the latter, strategic targeting of ISIS, in both Iraq and Syria, begins to make sense. Pinprick attacks while counting on "states" in the region to bear the brunt does not.

Afghanistan after NATO withdrawal is another obvious subject. Our study of alternate scenarios for Iraq post-US occupation could serve as a model. One can easily imagine three plausible, consequential, and distinctive futures for the country post-NATO withdrawal: a successful political consolidation, based either on power sharing or effective dictatorship, also premised on relative security; continued violence and struggle for power, but with external actors collectively avoiding intervention and containing the instability; and a regionalization of conflict that engulfed Pakistan, India, and China. Each alternate scenario would be assessed for relative probability and impact on US interests, and alternate US policies tested in and across the scenarios. If internal stability were deemed by far the best, but least likely, outcome, and regional spread most likely and most damaging to US interests, policy would be organized around containment. This might entail maintaining some US presence, to keep skin in the game; support for stability in Afghan politics even if this sacrificed constitutional procedures; and most likely ad hoc arrangements with Pakistan, India, China, and Russia to limit the spillover. Such collective action, even if informal, is admittedly a long shot, but if the spillover downside were accepted by all as both disastrous and most likely, a focused US effort at prevention might bring others along.

The future of the EU deserves serious scenario-like treatment. The multipolar, heterogeneous, globalized world I projected in chapter 1 is far more challenging to the EU than to any other major actor—a conclusion at odds with the immediate post–Cold War neorealist expectation that the EU could become America's next peer competitor. The multiplicity of membership, still immature EU institutions, limited popular legitimacy, tepid economic growth, increasing political extremism, dismal demographics, and diminishing external security all put the EU project in serious peril. New East-West fissures are now developing as members react disparately to Russian challenges. This adds to the older North-South division over macroeconomic policies and EU budget priorities. Something will have to give, and an alternate scenario study will help in identifying what it is, and what dangers and new sources of leverage this presents to the United States. There are very big issues at stake here, issues that will continue to challenge the Asian rebalancing of Obama and of any future president.

Obviously alternate futures for Ukraine, and for Central Europe generally, also need to be assessed, independently or, more usefully, as part of the EU study. Ukraine is being pushed by Russia inexorably toward the EU and NATO, and this will accelerate with Russian occupation of parts of eastern Ukraine. Yet the EU is already overexpanded, too dependent on Russian energy, and too deeply divided on many issues to offer Ukraine a track toward full membership. That leaves NATO, and a growing temptation to make an offer, with the risk of exacerbating East-West divisions inside NATO, and of a new cold war with Russia. Getting ahead of these issues will enable a fuller, more proactive consideration of US policy options in the presence of alternate Ukraine scenarios, from an all-out invasion and rapid deterioration in European security, to continued Russian intimidation and destabilization, to accommodation.

The future Asian balance of power is another obvious candidate for an alternate scenario project. China's rise and increasing assertiveness (of which a former Obama national security official said, "We didn't see this coming, and there's a lot of debate about how to counter it")[3] are the centerpiece of this, but the "realist pessimist" scenario is by no means the only possibility. For one thing, internal Chinese economic, demographic, and political developments will shape Chinese power and purpose, and big questions loom over the CCP's ability to shift the economy toward a more sustainable growth path, and if it does, the political changes this will set in motion. The foreign policies and internal political economies of other Asian actors are also in motion, adding to the complexity and uncertainty. Constructing plausible and useful scenarios for Asia is therefore much more than imagining "black boxes" engaged in alternate balance-of-power dynamics. Further complicating are the cross-cutting, often incompatible motivations operating on China, on other Asian actors, and on the United States. Very much unlike the Cold War is the combination of growing insecurity stemming from China's rise and from Japan's reaction on the one hand, and deep trade and financial interdependencies among all the actors, on the other. Getting the balance right, between containment and engagement, and doing this on a collective basis, with a diverse set of Asian allies, is a large challenge. How to maximize our interests in alternate, plausible Asian futures—how to encourage the future we want, how to deflect or manage the future we don't but may not have the power to avoid—would be the purpose of such an exercise.

The above ideas are primarily top-down, structured around existing approaches to American policy and strategy that are exposed to hidden risks, making these explicit, suggesting course corrections or fundamental changes when appropriate. But the record of post–Cold War grand strategy is decidedly ungrand, more variable and event driven, both within (Bush II) and between (Bush I to Clinton; Bush II to Obama; Obama to?) administrations. Given rapid change and the relative diminution of our power, thinking about the future of strategy requires that we get better at shaping strategy to emerging global trends that, in multipolarity, are less under our control. This is not to advocate passivity or neo-isolationism, but to suggest that a more acute understanding of the world as it's becoming, is indispensable to an effective use of our still considerable power.

Such studies should begin with a look at alternate futures for the overall global system. I've suggested multipolarity, conflicting identities, and illiberal globalization as comprising one, and in my view most likely, future; and the policy implications consisting of reducing our investment in primacy and liberal order building, and using our power robustly when our primary regional and global interests are at stake. Others will see the driving forces and most likely outcomes differently. A scenario study along these lines would give persuasive advocates of alternate systemic paradigms an opportunity to make the best case they can, interactively and within the discipline imposed by the alternate scenario process. It would also bring together theory-driven arguments about the future, with examination of factors "outside" these models, from technology change to demographic shifts, environmental and resource scarcities, and ideational transformation. A good place to start such an effort would be the NIC's every-five-year, long-term study of global trends, which includes alternate scenarios; but building scenarios would take the process to where the NIC cannot tread, into examination of US strategy, both as a driver of future change and in response to global forces that will challenge all actors.

The global economy has experienced nearly a decade of negative or slow growth, and the political costs for both prosperous and poor countries are self-evident. The duration of this growth crisis suggests the possibility of a "new normal" of slower growth resulting from political immobility (the United States, the EU), a decisive turn away from markets and interdependence (Russia), rising protectionism (India), and shifting sources of growth toward internal demand (China). Global trade liberalization has ground to a halt. This is plausible and consequential enough to be accepted as a premise for thinking about the global system, the limits to collaboration, the nature of future threats, the impact on the global balance of power (some states will do far better than others in such an environment), and US policy options.

One focus of such a study would be on emerging markets, that collection of countries celebrated pre-financial crisis as centers of growth, financial opportunity, and increasing power and leverage over the global system. Their weaknesses are now more worrisome than their strengths: stagnant economic growth, high levels of dollar-denominated debt (both corporate and, ultimately, sovereign), dwindling financial reserves, weakening currencies, high foreign participation in debt markets, and growing political risk. Countries with some or all of these characteristics include several that are pivotal: Russia, Ukraine, Turkey, Brazil, Argentina, Nigeria, Indonesia. The expected rise in US interest rates and further dollar strengthening makes some of these countries vulnerable to currency and banking crises and will pose threats to a still anemic global recovery, and to US strategic interests. It may also accelerate the rise of China as a global economic and financial player, as its huge reserves are placed at the disposal of financially strapped countries (including Russia).

The world I've projected produces big problems, mostly resulting from globalization and from weak collaboration in addressing them. Emerging issues—of climate, access to the global commons, global crime, terrorism, infectious disease, cyberwar, global economic management—will require some combination of strong global institutions, effective ad hoc cooperation among states, and American leadership. None of these requirements are abundant in the emerging world. These problems are thus more likely to fester, and become added sources of conflict. The US will find itself torn between the desire to lead global efforts at problem-solving, and the growing evidence of diminishing returns to leadership and the imperative of joining the fray in defense of primary interests, while leaving the rest for others to solve, or not. This suggests the value of an alternate futures think about the growing list of global problems, and how to rationally distribute the dwindling supply of US political capital across them. Some may be vital, and both demanding and admitting of common approaches (terrorism, global crime, global infectious disease, WMD nonproliferation); some may be vital but beyond the capacity of states in the emerging world to manage collectively (the Arctic, cyber-conflict); some may be vital and amenable to global responses, but only with new approaches (global economic management, climate change).

## Influencing Policy Debates

Grand strategy, successful or not, is based on assumptions about the world, and about the direction of change in variables that shape the context for policy. These assumptions may relate to global shifts in relative power, the behavior of key allies and adversaries, the advance of economic globalization and technology innovation, resource and environmental stress, and demographic pressures in aging and youthful societies. In the ideal strategy process, assumptions about these and other important "drivers" are made explicit, subjected to rigorous analysis and debate, and tested against emergent reality. Strategy advocates argue their case based on the robustness of these assumptions and their resilience over time. In this process of "bounded rationality," the best ideas emerge, and as the ground shifts—as it inevitability will—under current strategy (and as the strategy begins to visibly degrade), debate is reignited, previously rejected or new ideas are brought into play, and, if necessary, strategy adjustments or wholesale innovations are made.

This ideal model has rarely been achieved. Even in the relatively stable period of the Cold War, effective leadership (the missile crisis), political assassinations, popular revolutions (against the shah), top-down reform spinning out of control (glasnost and perestroika) have produced surprises, and altered the course of history. Despite Marshall's warning, US policy has frequently been premised on distorted assumptions, or has clung to assumptions long after being overtaken by events. We have survived these mismatches between policy and unfolding reality in part because of our enormous advantages in power and influence, in part because our competitive politics imposes accountability for mistakes and incentivizes new approaches.

Going forward, however, we will be penalized more severely for mistaken assumptions and stubborn adherence to outdated convention. Unless we more closely approximate the model of bounded rationality, the frequency of surprise and the negative consequences for our interests will increase. The two obvious reasons for this heightened urgency are growing complexity and diminished relative power. Complexity multiplies potential surprises. Diminished relative power reduces the capacity to recover, or to seize the opportunities that surprise has historically afforded the country.[4]

I have argued that an essential tool in adapting successfully to the emerging world, and shaping it when we can, is mastery of the alternate futures process. Part of such a process involves quick and accurate recognition of significant, policy-forcing shifts in the world (bottom-up studies), and part involves awareness of the downsides of prior strategic choices that could, unless managed, undermine existing strategy (top-down studies). Such a mentality, of prudential alertness, requires anticipatory scenario studies of the sort suggested earlier in this chapter. The identification of such studies itself requires a sense of where the big, impending changes are coming from, and where the built-in vulnerabilities of current policy reside. Such studies will have maximum policy effect if conceived within an already well-developed sense of future risk. Study outcomes can then be shaped to highlight those areas where extant assumptions are shaky, where policy will produce diminishing returns or perverse results unless adjusted to new developments, or where forces not anticipated by current thinking will require a new paradigm.

Policy impacts of scenario studies are also a function of who participates. For a multitude of reasons, policymakers should be at the table, fully engaged in selecting scenario ideas for in-depth study, discussing drivers of change and potential wild cards, and of course working through policy implications. Policymakers have subject matter expertise valuable in discussing drivers and will know the current intelligence on the country or issue under study. They are by definition sensitive to the types of study outcomes that would be considered consequential for current and future policy and strategy. They also know the policy process and can help shape outcomes to maximum impact. They can help to promote the results, arranging high-level briefings. Their names on the document reinforce credibility.

But realizing the full value of alternate scenarios requires more than ad hoc studies, important as they are for establishing baseline alternate futures to track change and test alternate policies. We should try to go beyond exploiting "policy windows," occasioned by new administrations, victory in war, or big surprises. While such demands are important opportunities to do useful work and expand the constituency for alternate futures, they are episodic, while the need for "what if" thinking is continuous, because change is continuous and risk is ever-present. Even when strategy appears to be succeeding, downsides must be monitored and assessed and responses rehearsed.

This calls for an ongoing process, owned by those responsible for global strategy, deriving legitimacy from support at the top, with enough credibility to deliver bad news even when it's least welcome or appears unjustified by events. A commitment to process attracts participants from policy agencies, and from top academic institutions, thus improving quality of the output, in terms of both rigor and policy relevance. It raises the state of the art of scenario building, thus improving quality over time. It legitimizes the approach and its practitioners, thus inculcating and protecting what-if, independent thinking, even when it's wrong. Most important, process continuously exposes policymakers to the analysis of ongoing change and consequences for existing policy and strategy. As Aaron Friedberg, deputy assistant for national security in the vice president's office from 2003 to 2005 wrote, "The true aim of national strategic planning is heuristic; it is an aid to collective thinking at the highest echelons of government, rather than a mechanism for the production of operational plans. The point is nicely captured in President Eisenhower's pithy observation that whereas 'plans are useless ... planning is indispensable.'"[5] And Peter Feaver, from 2005 to 2007 a special adviser to the NSC on strategic planning, described "what we learned ... [A]lthough several of our initiatives, including planning sessions, studies and 'second look' memos either posited scenarios that never came to pass or never resulted in concrete action, the very process of doing them enhanced the NSC's strategic outlook and helped prepare ... White House principals to respond better to other contingencies."[6]

In its combination of subject matter expertise and policy influence and savvy, its devotion to rigor in scenario building, and its continuous functioning, an institutionalized alternate futures process is uniquely capable of informing policy in the world as it's becoming. The specifics of how to organize such an effort will vary with administration, its key officials, and their styles, personal relationships, policy priorities. What is essential is an ongoing commitment to scrubbing assumptions, evaluating strategy effects, creating alternate scenarios as extant wisdom begins to fray, considering policy changes to revitalize current strategy, or entertaining new strategies. Some of the baseline studies recommended earlier may hold up nicely as templates for evaluating risk and thinking through alternate approaches. Events, however, may degrade the value of such studies, suddenly or over time. They should be re-examined periodically, either by on-site convocations of the original scenario construction group, or via online discussions. In addition, any policy official, having observed an event or trend that seems to challenge assumptions underlying current policy, should be able to quickly assemble a scenario group of the highest quality—from in and outside government—and conduct a structured brainstorm of the sort described in chapter 4, but without the long analytic preparation, process debates, reiteration of basic purpose, and expert introductions that consume so much time in ad hoc studies.

One imperative: the process must be led from the top, most likely from the National Security Council, which oversees the national security strategy process, seems convinced of the value of alternate futures, and has direct access to the president. This will require that someone or some office at NSC be mostly removed from day-to-day operational activities, which in turn requires that the value of the alternate futures process be internalized at NSC, legitimized, and rewarded, and that ongoing policy debates on immediate issues welcome inputs from strategic planning. Robert Bowie, Eisenhower's director of policy planning at State, remarked that "if insights and thinking on long-term factors are to be effective, they must be brought to bear on such decisions as they are made."[7]

Only top-level oversight and direct participation can confer legitimacy, attract the best experts, and maximize the value of the approach, while avoiding the disabling pitfalls described in chapter 4. Executive branch agencies must of course fully contribute their expertise and point of view, most notably the Policy Planning Bureau at State and Policy Planning at Treasury. The National Intelligence Council, with its scenario experience and access to worldwide expertise, might be a useful place to organize communities of experts to contribute to scenario building. Much of the administrative load could be carried by an outside contractor. But the overall project, its design, the identification of key questions, keeping the value proposition in view, the shape of its output, and the analysis of policy implications should be under the control of the NSC, not subcontracted to another agency or outside contractor.

# NOTES

## Introduction

1. Dean Acheson, *Present at the Creation: My Years in the State Department* (New York: W.W. Norton, 1969), 214.

2. James Steinberg, quoted in Bruce W. Jentleson, "An Integrative Executive Branch Approach to Policy Planning," in *Avoiding Trivia: The Role of Strategic Planning in American Foreign Policy*, ed. Daniel W. Drezner (Washington, D.C.: Brookings Institution, 2009), 78.

## Chapter 1

1. Michael F. Oppenheimer, "From Prediction to Recognition: Using Alternate Scenarios to Improve Foreign Policy Decisions," *SAIS Review of International Affairs* 32, no. 1 (Spring 2012): 19–31. MO note: I've said in the acknowledgments that portions of this and other chapters come from this article, which SAIS has given me written permission to use. Do we have to footnote each use?

2. Richard K. Betts, *Conflict after the Cold War: Arguments on Causes of War and Peace* (Pearson Education, 2013), 105–106.

3. Betts, *Conflict after the Cold War*, 53.

4. Scenarios Initiative, NYU Center for Global Affairs, cgascenarios.wordpress.com.

5. Adam Posen, "It's Still Coming," Peterson Institute for International Economics, June 5, 2009.

6. "Total Official: Europe Could Up LNG Purchases to Bypass Russia, Algeria," *Moscow Times,* June 9, 2010.

7. "Russia Energy Min. Sees South Stream Cost at $20 Bln, Reuters," Reuters.co.uk, July 30, 2008.

8. Terry Macalister, "Putin Attacks GM for Opel and Vauxhall U-turn," Guardian.co.uk, November 5, 2009.

9. "Turkish Contracting in the International Market," *Belga Yatirim,* March 12, 2010.

10. Alexander Zaitchik, "Russia Pours Billions in Oil Profits into Nanotech Race," *Wired News,* November 2, 2007.

11. Ellen Barry, "Research Group Urges Radical Changes in Russia," *New York Times,* February 3, 2010.

12. "Medvedev Throws Down the Gauntlet," *Russia Monitor,* September 11, 2009.

13. Luke Harding, "Russia Fears Embrace of Giant Eastern Neighbour, China," Guardian.co.uk, August 2, 2009.

14. Vladimir Socor, "Gazprom Executive Confirms Production and Investment Woes," *European Dialogue,* June 17, 2009.

15. "Putin's United Russia Party Punished at the Ballot Box," *Deutsche Welle,* March 15, 2010, http://www.dw-world.de.

16. "China, Russia, Sign 12 Agreements during Putin's Visit," *English Xinhua,* July 30, 2010.

17. "Timber Industry 2000–2004," *Kommersant—Russia's Daily,* http://www.kommersant.com/.

18. "Russia Steel Production Has a Suicide Look," *Asia Times Online,* March 31, 2009.

19. Tuomas Forseberg and Graeme P. Herd, "Divided West, European Security and the Transatlantic Relationship," Chatham House-Blackwell 2006, 113.

20. "Gazprom-Naftogaz Merger Positive?" UPI On-Line, May 5, 2010, http://www.upi.com/.

21. Anatoly Medetsky, "Medvedev Lands $20 Bln Nuclear Deal in Turkey," *Moscow Times,* May 13, 2010.

22. "Would Russians in Ferghana Valley Guarantee Stability or Spell Disaster?" Radio Free Europe / Radio Liberty, August 18, 2009.

23. "Russia and Ukraine Improve Soured Relations—Russian President," *RIA Novosti*, May 16, 2010.

24. Max Bergmann, "Germany's Russia Moment," *News World Politics Review*, National Security Network, May 19, 2009, 3.

25. "ICDA-CIS Gaining Ferro-alloys Market Share," May 27, 2010, http://www.metal-pages.com/news/story/47000/.

26. "High-Speed Train Service among Russia's Priorities-Putin," Voice of Russia, February 27, 2010, http://english.ruvr.ru.

27. Max Bergmann and Adomeit Hannes, "Germany's Policy on Russia: End of the Honeymoon?" IFRI, September 2005, 3.

28. Scenarios Initiative, "China 2020," NYU Center for Global Affairs, March 4, 2010, cgascenarios.wordpress.com.

29. Keith Bradsher, "An Opportunity for Wall St. in China's Surveillance Boom," *New York Times*, September 11, 2007, accessed November 28, 2009, http://www.nytimes.com/2007/09/11/business/worldbusiness/11security.html?ei=5088&en=12c1173fd8e333b5&ex=134 7163200&partner=rssnyt&emc=rss&pagewanted.

30. Address by Premier Wen Jiabao at the Second Western China International Cooperation Forum, October 16, 2009, accessed November 29, 2009, "Towards Greater Development and Opening-up of Western China," http://www.fmprc.gov.cn/eng/zxxx/t621406.htm.

31. "International Organization for Standardization," http://www.iso.org/iso/home.htm.

32. Keith Bradsher, "An Opportunity for Wall St. in China's Surveillance Boom," *New York Times*, September 11, 2007, accessed November 28, 2009, http://www.nytimes.com/2007/09/11/business/worldbusiness/11security.html?ei=5088&en=12c1173fd8e333b5&ex=134 7163200&partner=rssnyt&emc=rss&pagewanted.

33. Daniel Drezner, ed., *Avoiding Trivia: The Role of Strategic Planning in American Foreign Policy* (Washington, DC: Brookings Institution Press, 2009), 64.

34. R. Jervis, M. Roskin, P. E. Tetlock, C. B. McGuire, Y. F. Khong, Winter, et al., in *American Foreign Policy*, 5th ed., ed. G. Ikenberry (New York: Pearson Longman, 2005).

35. Richard K. Herrmann and Jong Kun Choi, "From Prediction to Learning: Opening Experts' Minds to Unfolding History," *International Security* 31 (2007): 132–161, doi:10.1162/isec.2007.31.4.132.

36. Michael Fitzsimmons, "The Problem of Uncertainty in Strategic Planning," *Survival: Global Politics and Strategy* 48 (2006): 131–146.

37. Roger C. Altman, "The Great Crash, 2008," *Foreign Affairs*, January–February 2009.

38. Scott Baldauf, "With Crocker's Exit, a Chance for a New Approach to Afghanistan," *Christian Science Monitor*, May 23, 2012.

## Chapter 2

1. Richard Clark, *Against All Enemies: Inside America's War on Terror* (New York: Free Press, 2004).

2. Richard A. Clark, "Against All Enemies", Free Press, New York, 2004, pps. 238–239.

3. Michael Fitzsimmons, "The Problem of Uncertainty in Strategic Planning," *Survival: Global Politics and Strategy* 48 (2006): 131–146.

4. F. Gregory Gause III, "Why Middle East Studies Missed the Arab Spring," *Foreign Affairs*, July–August 2011.

5. Gause, "Middle East Studies," 87.

6. Gause, "Middle East Studies," 90.

7. Nassim Nicholas Taleb and Mark Blyth, "The Black Swan of Cairo," *Foreign Affairs*, May–June 2011.

## Chapter 3

1. Michael Fitzsimmons, "The Problem of Uncertainty in Strategic Planning," *Survival: Global Politics and Strategy* 48 (2006): 131–146.

2. Francis Fukuyama, ed., *Blindside: How to Anticipate Forcing Events and Wild Cards in Global Politics* (Baltimore: Brookings Institution, 2007).

3. Roger C. Altman, "The Great Crash, 2008," *Foreign Affairs*, January–February 2009.

4. Richard K. Herrmann and Jong Kun Choi, "From Prediction to Learning: Opening Experts' Minds to Unfolding History," *International Security* 31 (2007): 132–161, doi:10.1162/isec.2007.31.4.132.

5. Peter Schwartz and Doug Randall, "Ahead of the Curve: Anticipating Strategic Surprise," in Fukuyama, *Blindside*, 105–6.

6. Andrew Erdmann, "Foreign Policy Planning through a Private Sector Lens," in *Avoiding Trivia: The Role of Strategic Planning in American Foreign Policy*, ed. Daniel W. Drezner (Washington, D.C.: Brookings Institution, 2009), 140.

7. Erdmann, "Foreign Policy Planning," 144.

8. Gregory F. Treverton and Jeremy J. Ghez, *Making Strategic Analysis Matter* (Santa Monica, Calif.: RAND, National Security Research Division, 2012), 13.

9. Nate Silver, *The Signal and the Noise* (New York: Penguin, 2012), 45.

10. Philip E. Tetlock, *Expert Political Judgement* (Princeton: Princeton University Press, 2009).

11. Tetlock, *Expert Political Judgment*, 85.

12. Tetlock, *Expert Political Judgment*, 21.

13. Tetlock, *Expert Political Judgment*, 118.

14. Tetlock, *Expert Political Judgment*, 119.

15. Tetlock, *Expert Political Judgment*, 101.

16. Tetlock, *Expert Political Judgment*, 113.

17. Tetlock, *Expert Political Judgment*, 104.

18. Richard N. Haass, "Planning for Policy Planning," in Drezner, *Avoiding Trivia*, 106.

19. Haass, "Planning for Policy Planning," 25.

20. G. John Ikenberry, *After Victory: Institutions, Strategic Restraint, and the Rebuilding of Order after Major Wars* (Princeton: Princeton University Press, 2001).

21. Jeffrey Legro, *Rethinking the World: Great Power Strategies and International Order* (Ithaca, N.Y.: Cornell University Press, 2005).

22. Robert Lempert, "Can Scenarios Help Policymakers Be Both Bold and Careful?" in Fukuyama, *Blindside*, 116.

## Chapter 4

1. "Scenarios Initiative," NYU Center for Global Affairs, cgascenarios.word-press.com.

2. Minxin Pei, *China's Trapped Transition: The Limits of Developmental Autocracy* (Cambridge, Mass.: Harvard University Press, 2005).

3. Bruce Gilley, *China's Democratic Future: How It Will Happen and Where It Will Lead* (New York: Cambridge University Press, 2004).

4. Randall Peerenboom, *China Modernizes: Threat to the West or Model for the Rest?* (New York: Oxford University Press, 2007).

5. Jianwu He and Louis Kuijs, "Rebalancing China's Economy: Modeling a Policy Package," World Bank China Research Paper No. 7, 2007, accessed

September 26, 2009, http://www.worldbank.org.cn/english/content/
working_paper7.pdf.

6. Fred C. Bergsten, Charles Freeman, Nicholas R. Lardy, and Derek
J. Mitchell, *China's Rise: Challenges and Opportunities* (Washington,
D.C.: Peterson Institute, 2008), 105.

7. He and Kuijs, "Rebalancing China's Economy."

8. Jamil Anderlini, "Rule of the Iron Rooster," *Financial Times*, August 25,
2009, 5.

9. Ralph Atkins, "China Edges Ahead of Germany in Race to Be 'World
Export Champion,'" *Financial Times*, August 25, 2009, 1.

10. Yongding Yu, "China Needs to Stimulate Reform, Not Only the
Economy," *Financial Times*, August 26, 2009, 7.

11. Anderlini, "Iron Rooster," 5.

12. Yu, "China Needs to Stimulate," 7.

13. Bergsten et al., *China's Rise*, chapter 6.

14. He and Kuijs, "Rebalancing China's Economy."

15. Hyun Jin Choi, "Fueling Crisis or Cooperation? The Geo-politics of
Energy Security in Northeast Asia," *Asian Affairs: An American Review*,
March 22, 2009.

16. Choi, "Fueling Crisis or Cooperation?"

17. Jing Fu, "Emissions Target Set for Govt Schemes," *China Daily*, June
6, 2009, accessed August 28, 2009, http://www.chinadaily.com.cn/
world/2009green/2009-06/06/content_8256019.htm.

18. Andrew Light and Julian L. Wong, "Statement on Today's UN Climate
Change Summit," September 22, 2009, accessed November 28, 2009, http://
www.americanprogress.org/pressroom/statements/2009/09/unspeech.Html.

19. Elizabeth Economy, Testimony before the Senate Foreign Relations
Committee. Hearing on "Challenges and Opportunities for U.S.-China
Cooperation on Climate Change," June 4, 2009, accessed on August 28, 2009,
http://www.cfr.org/publication/19570/prepared_testimony_on_challenges_
and_opportunities_for_uschina_cooperation_on_climate_change.html.

20. David Dollar, "Poverty, Inequality, and Social Disparities during China's
Economic Reform," World Bank Policy Research Working Paper No. WPS
4253, June 2007, accessed August 28, 2009, http://www-wds.worldbank.org/
servlet/WDSContentServer/WDSP/IB/2007/06/13/000016406_20070
613095018/Rendered/PDF/wps4253.pdf.

21. He and Kuijs, "Rebalancing China's Economy."

22. Dollar, "Poverty, Inequality."

23. Michael Anti, "China's Millions of Jobless Migrants," *World Policy Journal* 26, no. 1 (Spring 2009): 27–32.

24. "National Minorities Policy and Its Practice in China," Chinese Government Whitepaper, 2004, accessed August 28, 2009, http://www.fmprc.gov.cn/ce/ceee/eng/ztlm/zfbps/t112922.htm.

25. Dru Gladney, "China's Ethnic Tinderbox," BBC News, July 9, 2009, accessed August 28, 2009, http://news.bbc.co.uk/2/hi/asia-pacific/8141867.stm.

26. Gladney, "China's Ethnic Tinderbox."

27. Chuanjiao Xie, "Number of Seminaries on Increase in China," *China Daily*, October 22 2008, accessed August 29, 2009, http://www.Chinadaily.com.cn/china/2008-10/22/content_7127088.htm.

28. Xie, "Number of Seminaries."

29. "National Minorities Policy."

30. Gladney, "China's Ethnic Tinderbox."

31. All demographics information used in this section comes from the "Allianz Knowledge Report: Demographics Profile China—People Power," August 25, 2008, http://knowledge.allianz.com/en/globalissues/demographic_profiles/china/china_population_demographics.html.

32. Clement Chu S. Lau, "The Role of NGOs in China: A Critique of Conventional Wisdom," *Quarterly Journal of Ideology* 31, nos. 3–4 (2008), accessed August 29, 2009, http://www.lsus.edu/la/journals/ideology/contents/vol32/NGOs in China article 2008.8.pdf.

33. Zhang Ye, "China's Emerging Civil Society," Brookings Institution, August 2003, accessed August 29, 2009, http://www.brookings.edu/papers/2003/08china_ye.aspx.

34. Lau, "Role of NGOs."

35. Wing Thye Woo, "China's Short-Term and Long-Term Economic Goals and Prospects," Brookings Institution, http://www.brookings.edu/testimony/2009/0217_chinas_economy_woo.

36. "Total Number of Chinese Internet Users Reaches 338 million," *China News Wrap*, July 16, 2009, http://chinanewswrap.com/2009/07/16/total-number-of-chinese-Internet-users-reaches-338-million/.

37. Ashley Esarey and Qiang Xiao, "Political Expression in the Chinese Blogosphere: Below the Radar," *Asian Survey* 48, no. 5 (September–October 2008): 752–772, accessed August 27, 2009, http://caliber.ucpress.net/doi/abs/10.1525/AS.2008.48.5.752?cookieSet=1&journalCode=as.

38. Esarey and Xiao, "Political Expression."

39. Esarey and Xiao, "Political Expression."

40. Bergsten et al., *China's Rise*, 105.

41. Cheng Li, "China's Team of Rivals," *Foreign Policy*, March–April 2009, 88–93, accessed August 30, 2009, http://www.foreignpolicy.com/story/cms.php?story_id=4686.

42. Serhy Yekelchyk, *Ukraine: Birth of a Modern Nation* (Oxford: Oxford University Press, 2007), 4.

43. Andrew Wilson, *The Ukrainians: Unexpected Nation* (New Haven: Yale University Press, 2009), 172.

44. Paul J. D'Anieri, *Understanding Ukrainian Politics: Power, Politics, and Institutional Design* (Armonk, N.Y.: M.E. Sharpe, 2007), 49.

45. "Ukraine," *Freedom in the World 2010*, Freedom House, May 2010.

46. Grzegorz Gromadzki, Veronika Movchan, Mykola Riabchuk, Solonenko Iryna, Susan Stewart, Oleksandr Sushko, and Kataryna Wolczuk, *Beyond Colours: Assets and Liabilities of "Post-Orange" Ukraine* (Kyiv: International Renaissance Foundation; Warsaw: Stefan Batory Foundation, 2010), 48, 89.

47. Gromadzki et al., *Beyond Colours*, 46–47.

48. Gromadzki et al., *Beyond Colours*, 46.

49. Gromadzki et al., *Beyond Colours*, 46–47.

50. James Sherr, *The Mortgaging of Ukraine's Independence*, Report No. REP BP 2010/01 (London: Chatham House Russia and Eurasta Programme, August 2010), 2–6.

51. "Ukraine Rules in Favour of Stronger Presidential Rule," *Reuters*, October 1, 2010, accessed October 4, 2010.

52. D'Anieri, *Understanding Ukrainian Politics*, 54.

53. D'Anieri, *Understanding Ukrainian Politics*, 148–149; Janusz Bugajski, Steven Pifer, Keith Smith, and Celeste A. Wallandar, *Ukraine: A Net Assessment of 16 Years of Independence* (Washington, D.C.: Center for Strategic & International Studies, February 2008), 14.

54. D'Anieri, *Understanding Ukrainian Politics*, 148–149; Bugajski et al., *Ukraine*, 5.

55. D'Anieri, *Understanding Ukrainian Politics*, 148–149; Bugajski et al., *Ukraine*, 14.

56. D'Anieri, *Understanding Ukrainian Politics*, 51.

57. D'Anieri, *Understanding Ukrainian Politics*, 52.

58. Bugajski et al., *Ukraine*, 15.

59. Grzegorz et al., *Beyond Colours*, 32.

60. Bugajski et al., *Ukraine*, 15.

61. Bugajski et al., *Ukraine*, 4.

62. Sherr, *Mortgaging of Ukraine's Independence*, 5.

63. D'Anieri, *Understanding Ukrainian Politics*, 61.

64. Gromadzki et al., *Beyond Colours*, 26, 64.

65. "Ukraine," CIA World Factbook, 2010.

66. Steven Pifer, *Averting Crisis in Ukraine*, Special Report No. 41 (New York: Council on Foreign Relations, January 2009), 20.

67. Mykola Riabchuk, "Ambivalence or Ambiguity? Why Ukraine Is Trapped between East and West," in *Ukraine, the EU and Russia: History, Culture and International Relations*, ed. Stephen Velychenko (Basingstoke: Palgrave Macmillan, 2007), 80.

68. Wilson, *The Ukrainians*, 220, 233, 251–252.

69. Yekelchyk, *Ukraine*, 228.

70. Pifer, *Averting Crisis in Ukraine*, 16.

71. Wilson, *The Ukrainians*, 217.

72. Riabchuk, "Ambivalence or Ambiguity," 80.

73. Riabchuk, "Ambivalence or Ambiguity," 80.

74. Taras Kuzio quoted in Olearchyk Roman, "Unbalanced Between East and West," *Financial Times*, June 1, 2010.

75. Yekelchyk, *Ukraine*, 4; Alexander J. Motyl, "Can Ukraine Have a History?" *Problems of Post-Communism* 57, no. 3 (May–June 2010): 51–61.

76. Proc. of Ukraine: A Net Assessment of 16 Years of Independence Implementation of Key Tasks and Recommendations, March 6, 2009.

77. Motyl, "Can Ukraine Have a History?" 55.

78. Taras Kuzio, "Shift to Soviet-Russian National Identity in Ukraine," *Eurasia Daily Monitor* 7, no. 90 (September 16, 2010).

79. Taras Kuzio, "Ukraine: Leaving the Crossroads," in *Central and East European Politics: From Communism to Democracy*, ed. Sharon L. Wolchik and Jane L. Curry (Lanham, MD: Rowman & Littlefield, 2008), 365.

80. Pifer, *Averting Crisis in Ukraine*, 16.

81. Bugajski et al., *Ukraine*, 5, 16.

82. Anders Åslund and Oleksandr Paskhaver, *Proposals for Ukraine: 2010—Time for Reform* (Kyiv: International Experts Commission, 2010), 11.

83. Åslund and Paskhaver, *Proposals for Ukraine*, 27–29.

84. Bugajski et al., *Ukraine*, 16–17; Jan Neutze and Adrian Karatnycky, *Corruption, Democracy, and Investment in Ukraine* (Washington, D.C.: Atlantic Council of the United States, October 2007), 10.

85. Bugajski et al., *Ukraine*, 17.

86. *World Economic Outlook 2010*, database, International Monetary Fund, http://www.imf.org/external/pubs/ft/weo/2010/02/weodata/index.aspx.

87. "Ukraine," CIA World Factbook, 2010.

88. Samuel Charap, "Seeing Orange," *Foreign Policy*, January 18, 2010.

89. Bugajski et al., *Ukraine*, 16.

90. Åslund and Paskhaver, *Proposals for Ukraine*, 10–13.

91. "Ukraine," European Trade Commission, http://ec.europa.edu/trade/creating-opportunities/bilateral-relations/countries/ukraine/.

92. Calculated from statistics on Ukraine from the European Trade Commission. (http://trade.ec.europa.eu/doclib/docs/2006/september/tradeoc_113459.pdf); "Ukraine," CIA World Factbook, 2010.

93. Bugajski et al., *Ukraine*, 20–21.

94. Åslund and Paskhaver, *Proposals for Ukraine*, 45.

95. Kirsten Westphal, "Russian Gas, Ukrainian Pipelines, and European Supply Security," SWP German Institute for International and Security Affairs, September 2009, 7.

96. "The Russian-Ukrainian Gas Trade," *Russian Analytical Digest* 75 (March 16, 2010): 10.

97. Steven Pifer, "Ukraine's Geopolitical Choice, 2009," *Eurasian Geography and Economics* 50, no. 4 (2009): 395.

98. Westphal, "Russian Gas," 27.

99. Edward C. Chow, "Neighborly Corporate Raid," Center for Strategic and International Studies, May 7, 2010; Michael Emerson, "Time for a Tripartite Gas Pipeline Consortium for Ukraine," CEPS Commentary, Centre for European Policy Studies, June 8, 2010.

100. Westphal, "Russian Gas," 14.

101. Chow, "Neighborly Corporate Raid."

102. Neutze and Karatnycky, *Corruption, Democracy*.

103. Alexandros Petersen and Tamerlan Vahabov, "Ukraine's Energy Reform Opportunity," *New Observer*, April 13, 2010.

104. Westphal, "Russian Gas," 12.

105. Åslund and Paskhaver, *Proposals for Ukraine*, 45.

106. Åslund and Paskhaver, *Proposals for Ukraine*, 46.

107. *An Avoidable Tragedy: Combatting Ukraine's Health Crisis*, Report no. 51829, World Bank, 2009.

108. Åslund and Paskhaver, *Proposals for Ukraine*, 55.

109. Åslund and Paskhaver, *Proposals for Ukraine*, 52.

110. Natalie Mychajlyszyn, "From Soviet Ukraine to the Orange Revolution," in *Europe's Last Frontier? Belarus, Moldova, and Ukraine between Russia and the European Union*, ed. Oliver Schmidtke and Serhy Yekelchyk (New York: Palgrave Macmillan, 2008), 31.

111. Mychajlyszyn, "From Soviet Ukraine," 31.

112. Pifer, "Ukraine's Geopolitical Choice, 2009," 13.

113. http://www.razumkov.org/ua/eng/poll.php?poll_id=305.

114. Bugajski et al., *Ukraine*, 18–19; Pifer, "Ukraine's Geopolitical Choice, 2009," 13.

115. Jeffrey Simon, "Ukraine against Herself: To Be Euro-Atlantic, Eurasian, or Neutral?" *Strategic Forum* 238 (February 2009): 2.

116. Gromadzki et al., *Beyond Colours*, 52.

117. Bugajski et al., *Ukraine*, 20–21.

118. Pifer, "Ukraine's Geopolitical Choice, 2009," 392–393.

119. Pifer, "Ukraine's Geopolitical Choice, 2009," 394.

120. Kuzio, "Ukraine: Leaving the Crossroads," 364.

121. Sherr, *Mortgaging of Ukraine's Independence*, 13.

122. Pifer, "Ukraine's Geopolitical Choice, 2009," 393.

123. Kuzio, "Ukraine: Leaving the Crossroads," 364.

124. Pifer, "Ukraine's Geopolitical Choice, 2009," 393.

125. Stefan Gänzle, "EU-Russia Relations and the Repercussions on the In-betweens," in Schmidtke and Yekelchyk, *Europe's Last Frontier?* 196.

126. Gänzle, "EU-Russia Relations," 196; and Pifer, "Ukraine's Geopolitical Choice, 2009," 394–395.

127. Kuzio, "Ukraine: Leaving the Crossroads," 364; and Pifer, "Ukraine's Geopolitical Choice, 2009," 395–396.

128. Adrian Karatnycky, "Orange Peels: Ukraine after Revolution," Atlantic Council, August 20, 2010.

129. Kuzio, "Ukraine: Leaving the Crossroads," 364.

130. Sherr, *Mortgaging of Ukraine's Independence.*

131. Arkady Moshes, "Ukraine between a Multivector Foreign Policy and Euro-Atlantic Integration," PONARS Policy Memo No. 426, December 2006, 1.

132. Francis Fukuyama, *Blindside: How to Anticipate Forcing Events and Wild Cards in Global Politics* (Baltimore: Brookings Institution, 2007), 94.

133. Scenarios Initiative, "Syria," NYU Center for Global Affairs, October 10, 2012, cgascenarios.wordpress.com.

134. John Ikenberry, "The Illusion of Geopolitics: The Enduring Power og the Liberal Order", Foreign Affairs, May/June, 2014.

135. Aaron Friedberg, "The Future of U.S.-China Relations: Is Conflict Inevitable?" *International Security* 30, no. 2 (2005): 7–45.

136. Friedberg, "U.S.-China Relations:," 39.

137. The actual results were 57.88 percent in favor to 42.12 percent against, with 73.71 percent of registered voters participating.

138. Bulent Aliriza and D. Koenhemsi, D. "Erdogan's Referendum Victory and Turkish Politics," CSIS Turkey Update, October 15, 2010, http://csis.org/files/publication/101510-Erdogan%27s-Referendum-Victory-and-Turkish-Politics.pdf.

139. Henri J. Barkey, "Turkey's New Global Role," interview, Carnegie Endowment for International Peace, November 17, 2010. http://www.carnegieendowment.org/ publications/index.cfm?fa=view&id= 41952.

140. Mustafa Turan, "Thousands Voice Support for Reform in Major Rally," Today's Zaman, August 30, 2010, http://www.todayszaman.com/news-220440-thousands-voice-support-for-reform-in-major-rally.html.

141. Omer Taspinar, "Judicial Independence and Democracy in Turkey," Brookings Institution, July 12, 2010, http://www.brookings.edu/opinions/2010/0712_turkey_democracy_taspinar.aspx.

142. "Coups Away," The Economist, February 11, 2010, http://www.economist.com/node/15505946.

143. http://www.washingtonpost.com/wp-dyn/content/article/2010/12/16/ AR2010121604908.html; "Next Hearing in Turkey's Ergenekon Case to Be Held in 2011," Hurriyet, November 12, 2010, http://www.hurriyetdailynews.com/n. php?n=next-ergenekon-hearing-in-2011-2010-11-12.

144. The new version of Article 145's first paragraph reads: "Military justice shall be exercised by military courts and military disciplinary courts. These courts shall only have jurisdiction to try military personnel for military offences related to military services and duties. Cases regarding crimes against the security of the State, constitutional order and its functioning shall be heard before the civil courts in any event." Draft Constitutional Amendments Proposal, Secretariat General for EU Affairs, translated by Secretariat General for European Union Affairs, effective March 20, 2010, Article 145, p. 17.

145. Cf. proposed changes in Article 145, as well as 125, among other articles; Draft Constitutional Amendments Proposal.

146. "In It For the Long Haul," The Economist, October 21, 2010, http:// www.economist.com/node/17276430.

147. As explained by Henri Barkey: http://latimesblogs.latimes.com/ babylonbeyond/2010/12/turkey-elections-akp-erdogan-parliament- opposition.html.

148. Playing the "Islamic card" imagines populist and media intensive statements and appearances of Prime Minister Erdogan, as seen, for example, at Davos in 2009 or, in the context of the conflict in Darfur, when he claimed that Muslims would not commit genocide; cf. Seth Freedman, "Erdogan's Blind Faith in Muslims: The Turkish Leader's Support of Sudan's Omar al-Bashir While Condemning Gaza 'War Crimes' Play to Fears on the Israeli Right," Guardian, November 11, 2009, http://www.guardian.co.uk/ commentisfree /2009/nov/11/erdogan-muslims-turkish-sudan-gaza/print.

149. Turkey Country Report, Economist Intelligence Unit, January 2011.

150. Barkey, "Turkey's New Global Role."

151. Cf. Sabrina Tavernise, "Turkish Group Wields Wit as Tool for Political Change," New York Times, July 22, 2007, http://www.nytimes. com/2007/07/22/world/europe/22turkey.html?_r=1.

152. See, for example, Turkey 2010 Progress Report, published by the European Commission, Brussels, November 9, 2010.

## Chapter 5

1. Jeffrey W. Legro, "A 'Return to Normalcy'? The Future of America's Internationalism," in *Avoiding Trivia: The Role of Strategic Planning in American Foreign Policy*, ed. Daniel W. Drezner (Washington, D.C.: Brookings Institution, 2009), 64.

2. Helene Cooper and Michael R. Gordon, "Iraqi Kurds Expand Autonomy as ISIS Reorders the Landscape," *New York Times*, August 29, 2014.

3. David E. Sanger, "Commitments on Three Fronts Test Obama's Foreign Policy," *New York Times*, September 3, 2014.

4. John Lewis Gaddis, *Surprise, Security, and the American Experience* (Cambridge, Mass.: Harvard University Press, 2005).

5. Aaron Friedberg, "The Future of U.S. China Relations: Is Conflict Inevitable?" *International Security* 30, no. 2 (2005): 85.

6. Peter Feaver and William Inboden, "A Strategic Planning Cell on National Security at the White House," in Drezner, *Avoiding Trivia*, 106.

7. Drezner, *Avoiding Trivia*, 72.

# BIBLIOGRAPHY AND
# RECOMMENDED READINGS

Brands, Hal. *What Good Is Grand Strategy?* Ithaca, N.Y.: Cornell University Press, 2014.

Center for Global Affairs, Scenario Initiative, School of Professional Studies, NYU. Reports on the future of Iran, Iraq, China, Russia, Ukraine, Turkey, Pakistan, and Syria, 2009–2013. Available at www.cgascenarios.wordpress.com.

Danzig, Richard. "Driving in the Dark: Ten Propositions about Prediction and National Security." Center for a New American Democracy, October 2011.

Davis, Jack. *Uncertainty, Surprise and Warning.* CIA Directorate of Intelligence, Product Evaluation Staff, 1996.

Drezner, Daniel, ed. *Avoiding Trivia: The Role of Strategic Planning in American Foreign Policy.* Washington, D.C.: Brookings Institution Press, 2009.

Erdmann, Andrew. "Foreign Policy Planning through a Private Sector Lens." In *Avoiding Trivia: The Role of Strategic Planning in American Foreign Policy*, ed. Daniel W. Drezner. Washington, D.C.: Brookings Institution, 2009.

Fitzsimmons, Michael. "The Problem of Uncertainty in Strategic Planning." *Survival: Global Politics and Strategy* 48, no. 4 (2006): 131–146.

Fukuyama, Francis, ed. *Blindside: How to Anticipate Forcing Events and Wild Cards in Global Politics.* Washington, D.C.: Brookings Institution, 2007.

Fukuyama, Francis. *The End of History and the Last Man.* New York: Free Press, 1992.

Gause, F. Gregory, III. "Why Middle East Studies Missed the Arab Spring." *Foreign Affairs*, July–August 2011.

Gilley, Bruce. *China's Democratic Future: How It Will Happen and Where It Will Lead.* New York: Cambridge University Press, 2004.

Haass, Richard N. "Planning for Policy Planning." In *Avoiding Trivia: The Role of Strategic Planning in American Foreign Policy,* ed. Daniel W. Drezner. Washington, D.C.: Brookings Institution, 2009.

Herrmann, Richard, and Jong Kun Choi. "From Prediction to Learning: Opening Experts' Minds to Unfolding History." *International Security* 31, no. 4 (Spring 2007): 132–161.

Huntington, Samuel. *The Clash of Civilizations and the Remaking of World Order.* New York: Simon and Schuster, 1996.

Ikenberry, John. *Liberal Leviathan: The Origins, Crisis, and Transformation of the American World Order.* Princeton: Princeton University Press, 2011.

Jentleson, Bruce W. "An Integrative Executive Branch Approach to Policy Planning." In *Avoiding Trivia: The Role of Strategic Planning in American Foreign Policy,* ed. Daniel W. Drezner. Washington, D.C.: Brookings Institution, 2009.

Kahneman, Daniel. *Thinking, Fast and Slow.* New York: Farrar, Straus and Giroux, 2011.

Kupchan, Charles. *No One's World.* New York: Oxford University Press, 2012.

Legro, Jeffrey. *Rethinking the World.* Ithaca, N.Y.: Cornell University Press, 2007.

National Intelligence Council. *Global Trends 2030,* Washington, D.C., 2013.

Oppenheimer, Michael. "From Prediction to Recognition: Using Alternate Scenarios to Improve Foreign Policy Decisions." *SAIS Review of International Affairs* 32, no. 1 (Spring 2012): 19–31.

Peerenboom, Randall. *China Modernizes: Threat to the West or Model for the Rest?* New York: Oxford University Press, 2007.

Pei, Minxin. *China's Trapped Transition: The Limits of Developmental Autocracy.* Cambridge, Mass.: Harvard University Press, 2005.

Schwartz, Peter. *The Art of the Long View.* New York: Crown Business, 1991.

Schwartz, Peter, and Doug Randall. "Ahead of the Curve: Anticipating Strategic Surprise." In *Blindside: How to Anticipate Forcing Events and Wild Cards in Global Politics,* ed. Francis Fukuyama. Washington, D.C.: Brookings Institution, 2007

Sestanovich, Stephen. *Maximalist: America in the World from Truman to Obama.* New York: Vintage, 2014.

Silver, Nate. *The Signal and the Noise: Why So Many Predictions Fail, but Some Don't.* New York: Penguin, 2012.

Taleb, Nassim Nicholas. *The Black Swan: The Impact of the Highly Improbable.* New York: Random House, 2007.

Tetlock, Philip. *Expert Political Judgment.* Princeton: Princeton University Press, 2005.

Treverton, Gregory. "Making Policy in the Shadow of the Future." RAND Occasional Paper.

Treverton, Gregory, and Jeremy J. Ghez. *Making Strategic Analysis Matter.* RAND Conference Proceedings, National Security Division. Santa Monica, Calif.: RAND, 2012.

# INDEX